高等学校系列教材

土力学(双语)

SOIL MECHANICS

王　滢　高　盟　杨旭旭　主编

中国建筑工业出版社

图书在版编目（CIP）数据

土力学 = SOIL MECHANICS / 王滢，高盟，杨旭旭主编. -- 北京：中国建筑工业出版社，2025. 4. --（高等学校系列教材）. -- ISBN 978-7-112-30967-2

Ⅰ. TU43

中国国家版本馆 CIP 数据核字第 20251HX090 号

本书以编者 10 余年的土力学双语教学、科研工作的丰富资料和实践经验为基础，参阅国内外大量文献资料编写而成。围绕土力学的基本概念、基本原理和基本方法，重点阐述土体的力学行为及其在岩土工程中的应用。本教材共 10 章，主要内容包括：绪论，土的组成及其物理性质，渗透性与渗流，土中应力，土的压缩性，地基变形，土的抗剪强度，土压力，地基承载力及边坡稳定性。并附有习题及答案、教学课件和教学大纲。本书在强调内容的系统性和完整性的同时，突出了前瞻性、拓展性、实用性和简洁性等特点。

本书可作为高等院校土木工程、地质工程和交通工程等专业本科生双语教材和相关国际工程的技术人员培训教材，还可供相关行业的设计、施工等各类技术人员参考。

为方便教学与学习，本书附有习题及答案、教学课件、教学大纲和双语在线课程等资源，有需要可联系本书作者王滢（邮箱：hopeywwgy@hotmail.com），或责任编辑梁瀛元（邮箱：1512879111@qq.com，电话：(010)58337277）。

责任编辑：杨　允　梁瀛元
责任校对：芦欣甜

高等学校系列教材
土力学（双语）
SOIL MECHANICS
王　滢　高　盟　杨旭旭　主编

*

中国建筑工业出版社出版、发行（北京海淀三里河路 9 号）
各地新华书店、建筑书店经销
北京红光制版公司制版
建工社（河北）印刷有限公司印刷

*

开本：787 毫米×1092 毫米　1/16　印张：13　字数：323 千字
2025 年 4 月第一版　2025 年 4 月第一次印刷
定价：**49.00** 元（赠教师课件）
ISBN 978-7-112-30967-2
(44479)

前　言

在“一带一路”“中国制造2025”“工业4.0”等建设布局背景及《华盛顿协议》要求下，工程教育改革应与国际标准接轨。土力学作为土木工程类专业的一门重要的专业核心课程，其双语教学对构建国际教育课程体系，培养专业扎实、精通英语的土木工程专业国际化人才，满足教育“新工科”发展需求具有重要意义。双语教材建设是双语教学的重要内容。目前有关土力学的双语教材较少，而且很难满足“学生易学、教师易教和英文纯正”的教材要求。编者长期从事土力学双语教学，主持多项省级、校级土力学双语在线课程和教学改革研究项目，形成了较为系统的教学成果。融合双语教学过程中形成的教学理念，编者希望编写一本以双语为特色，系统、完整、简洁易懂，同时兼顾先进性、启发性和实用性的简明教材，为土木工程、交通工程、地质工程专业认证国际标准化以及培养国际化人才服务。

本教材共10章，主要包含3部分的内容：土及土力学的基本概念，包括绪论、土的组成及其物理性质，这部分内容是分析土力学问题的基础；土力学的基本理论，包括渗透性与渗流、土中应力、土的压缩性以及地基变形、土的抗剪强度，这部分内容是土力学的主线和核心内容；土力学的理论应用，包括土压力、地基承载力及边坡稳定性，这部分是土力学理论与工程应用的密切联系，是土力学的重要内容。本书第2章、第4章、第5章、第6章、第7章和第10章由王滢执笔；第1章、第3章、第8章和第9章由高盟、杨旭旭执笔。全书由王滢负责统稿，中国石油大学李大勇教授主审。

本书的编写和出版得到了山东省教学改革研究项目重点项目（Z2023144）的支持和帮助，在此表示感谢。感谢研究生李婉婷、刘嘉怡、耿源和隋浩唐等在整理、绘图等方面的帮助和付出。感谢中国建筑工业出版社编辑杨允、梁瀛元为本书出版所付出的辛勤劳动。此外，本书引用了国内外许多学者的研究成果和文献资料，在此，编者一并表示真挚的谢意。

王滢

2024年8月于青岛

目　　录

Chapter 1 Introduction

1.1 Concept and Importance of Soil Mechanics

Soil mechanics is a discipline that studies the physical and mechanical properties of soil, as well as the stress, deformation, strength, and seepage laws inside the soil under load, in order to solve problems such as deformation and stability of soil. It is a core course in civil engineering. It studies the mechanical behavior of soil under various loads, solves engineering problems, and is a discipline that combines theory and application very closely.

Common problems related to soil and geotechnical (marine) structures encountered in engineering include but are not limited to the following aspects:

(1) Physical and mechanical properties of soil;

(2) Stress, deformation, and strength characteristics of soil;

(3) Soil pressure issue;

(4) The bearing capacity problem and design of shallow foundation;

(5) Landslides of natural and artificial slopes;

(6) Design of pile foundation;

(7) Design of marine foundation;

(8) Stability of geotechnical (marine) structures.

(9) Physical, mechanical, and engineering properties of soil in weightlessness environment.

In order to conduct a reasonable analysis and design of these issues, it is necessary to have a deep understanding of the behavior of soil under complex load. For example, when designing foundations to resist loads in complex marine environments, engineering technicians must come up with special solutions based on local soil conditions and environmental factors. The basic design must meet the standards of static loads and be able to safely resist dynamic loads. When designing load conditions, engineering technicians are required to

第1章　绪论

1.1　土力学的概念及其重要性

土力学是研究土的物理力学性质以及在荷载作用下土体内部的应力、变形、强度和渗流规律，从而解决土体的变形和稳定等问题的一门学科，是土木工程专业的一门核心课程。它研究土体在各类荷载作用下的力学行为，解决工程问题，是一门理论与应用结合非常紧密的学科。

工程中经常遇到与土及土（海）工结构有关的问题，包括但不限于以下方面：

（1）土的物理力学性质；

（2）土的应力、变形和强度特性；

（3）土压力问题；

（4）浅基础的承载力问题和设计；

（5）天然和人工边坡的滑坡；

（6）桩基础的设计；

（7）海洋基础的设计；

（8）土（海）工结构的稳定性；

（9）失重环境下土的物理力学及工程性质。

为了对这些问题进行合理的分析和设计，必须深入理解土在复杂荷载下的行为。例如，在设计基础以抵抗海洋复杂环境下的荷载时，工程技术人员必须根据当地土质条件和环境因素得出特殊的解决方案。基础的设计必须满足静态荷载的标准，同时，必

answer the following questions:

(1) How to define the failure effect under complex ocean loads and what should be the failure criteria?

(2) What is the relationship between complex load effects and the geotechnical parameters used to define failure criteria?

(3) How to obtain marine geotechnical parameters?

(4) How to determine the acceptable safety factor under complex ocean loads, and whether the safety factor used under static design conditions is sufficient to ensure satisfactory performance, or whether additional conditions need to be met?

(5) How to obtain geotechnical parameters in weightlessness environment?

In recent years, with the vigorous development of infrastructure in China, a large number of new soil related issues have emerged, such as the stability of marine structures, the remediation of polluted soil, the physical and mechanical properties of seabed energy soil, and the physical and mechanical properties of lunar soil. The new engineering practice problems have posed new issues and challenges for soil mechanics. At the same time, these new problems in engineering construction have continuously injected new vitality into soil mechanics research. The importance of soil mechanics lies in providing safe, acceptable, and time-tested solutions for various engineering problems in soil, although some areas may lack information and actual load conditions may not be predictable, such as earthquakes and waves. In summary, it can be seen that soil mechanics is an interdisciplinary and complex field that requires an understanding of the mechanical behavior of soil under complex load, the principles and methods of mathematical mechanics, numerical methods such as finite element method, in order to find appropriate solutions to practical problems.

须能够安全地抵抗动态荷载。在设计荷载条件时，工程技术人员要回答以下问题：

（1）海洋复杂荷载作用下失效应如何定义，失效标准应是什么？

（2）复杂荷载作用与定义失效标准时使用的岩土参数之间的关系是什么？

（3）海洋岩土参数如何取得？

（4）海洋复杂荷载作用下可接受的安全系数如何确定，静态设计条件下使用的安全系数是否足以确保性能良好，或者是否需要满足附加条件？

（5）失重环境下岩土参数如何取得？

近年来，中国基础设施大力发展，与土有关的新问题大量涌现，例如，海工结构的稳定性，污染土的修复，海底能源土、月壤的物理力学性质等。新的工程实践问题为土力学提出了新课题、新挑战。同时，这些工程建设中的新问题也给土力学研究不断注入新的活力。土力学的重要性在于为土中的各类工程问题提供安全、可接受和能经得起时间检验的解决方案，尽管某些地区可能缺乏信息，并且实际荷载条件可能无法预测，例如地震、波浪。综上所述，土力学是一个跨学科的复杂领域，需要了解复杂荷载条件下土的力学行为、数学力学的原理和方法、有限元法等数值方法，以找到实际问题的适当解。

1.2 Main Research Contents of Soil Mechanics

The main contents of soil mechanics can be divided into three modules:

(1) The physical and mechanical properties of soil, including its formation, composition, three-phase ratio index, physical characteristics of cohesive and non-cohesive soils, engineering classification of soil, permeability, compressibility, etc., are the foundation of soil mechanics.

(2) The basic theories of soil mechanics, including Darcy's law, Boussinesq's fundamental solution, additional stress calculation formula, deformation calculation formula, effective stress principle, Terzaghi's one-dimensional consolidation theory, Coulomb's formula (one line, one circle), etc., are the core of soil mechanics.

(3) The engineering applications of soil mechanics include seepage calculation of foundations, settlement and consolidation of foundations, soil pressure problems, prediction of foundation bearing capacity, and analysis of slope stability. These are the key and difficult points of soil mechanics.

The ultimate goal of soil mechanics is to solve the problems of strength, deformation and stability of soil under various loads (force, temperature, etc.), that is, the third module problem. The research of the first module and the second module problem is to solve the third module problem.

1.2 土力学的主要研究内容

土力学的主要内容可分为三个模块：

（1）土的物理力学性质，包括土的形成、组成、三相比例指标、黏性土和无黏性土的物理特征、土的工程分类、渗透性、压缩性等，这部分内容是土力学的基础。

（2）土力学的基本理论，包括达西定律、布辛奈斯克基本解、附加应力计算公式、变形计算公式、有效应力原理、太沙基一维固结理论、库仑公式（一条线，一个圆）等，这部分内容是土力学的核心。

（3）土力学的工程应用，包括地基的渗流计算、地基的沉降与固结、土压力问题、地基承载力预测和边坡稳定性分析等，这部分内容是土力学的重点，也是难点。

土力学的最终目标是解决土体在各种作用下（力、温度等）的强度、变形和稳定性问题，即第3部分问题，第1部分和第2部分问题的研究都是为解决第3部分问题服务的。

1.3 Disciplinary Characteristics and Research Methods of Soil Mechanics

Soil mechanics is a discipline that studies soil, and soil is a complex material with natural variability, fragmentation, and three phases, which determines that soil mechanics is different from general mechanics such as material mechanics, theoretical mechanics, structural mechanics, and elastic mechanics, and has strong complexity and practicality.

1.3 土力学的学科特点及研究方法

土力学是研究土的一门力学，而土是一种复杂的材料，具有自然变异性、碎散性和三相性，决定了土力学不同于材料力学、理论力学、结构力学、弹性力学等一般力学，具有较强的复杂性和实践性。

Soil is a product of nature, formed by various types of sediments in the complex natural environment where rocks undergo physical, chemical, and biological weathering, as well as erosion, transport, and sedimentation processes. Therefore, the types of soil and their physical and mechanical properties vary greatly, with natural variability. Solid particles in soil are the detrital material of weathered rocks, abbreviated as soil particles. The soil particles form the skeleton of the soil, and the pores between the soil particles are filled with liquid water and gas. Therefore, the soil in nature is a three-phase mixture composed of soil particles (solid), water (liquid), and gas (gas). Soil is different from general continuous solids in that its particle size and shape are irregular, and there are pores between soil particles without a fixed shape. It is a typical discontinuous medium with fragmentation.

土是自然界的产物，是在复杂的自然环境中，岩石经历物理、化学、生物风化作用以及剥蚀、搬运、沉积作用等所形成的各类沉积物组成的。因此，土的类型及其物理、力学性状千差万别，具有自然变异性。土中固体颗粒是岩石风化后的碎屑物质，简称土粒。土粒构成了土的骨架，土粒与土粒之间的孔隙充填着液态水和气体。因此，自然界的土是由土粒（固体），土中水（液体）和土中气（气体）组成的三相混合物。土不同于一般的连续固体，土粒的大小、形状不规则，土粒与土粒间存在孔隙，没有固定的形状，是典型的非连续介质，具有碎散性。

Soil in engineering can be divided into general soil and special soil. General soil can be divided into inorganic soil and organic soil. Inorganic soil includes four categories: gravel soil, sandy soil, silt, and cohesive soil. Special soil includes collapsible soil, expansive soil, frost heave soil, red clay, soft soil, fill soil, mixed soil, saline soil, polluted soil, weathered rock, and residual soil, etc. There are many types of soil and their engineering properties are very complex. There are no such two locations on Earth where the engineering properties of soil are completely identical, which increases the difficulty of soil mechanics research. Soil is a discontinuous medium, and its stress-strain relationship exhibits nonlinearity and elastic-plastic behavior. The key to soil mechanics is to fully utilize indoor testing and testing techniques, establish constitutive models that are close to the essence of soil, and then use theoretical calculations and numerical simulation techniques to solve engineering problems related to soil.

工程中的土可分为一般土和特殊土。一般土又可分为无机土和有机土。无机土包括碎石土、砂土、粉土和黏性土四大类。特殊土有湿陷性土、膨胀土、冻胀土、红黏土、软土、填土、混合土、盐渍土、污染土、风化岩和残积土等。土的种类繁多，工程性质十分复杂，地球上不存在两个工程性质完全相同的地点，这为土力学的研究增加了难度。土是一种非连续介质，其应力-应变关系表现出非线性、弹塑性。土力学的关键是充分利用室内试验和测试技术，建立接近于土本质的本构模型，再利用理论计算和数值模拟技术解决与土有关的工程问题。

At present, there are three main ways and methods to solve soil mechanics problems: experiments, theoretical calculations, and numerical simulation analysis.

目前，解决土力学问题的主要途径和方法有3种：试验、理论计算、数值模拟分析。试验包

Experiments include indoor model tests, on-site tests, and in-situ testing measurements, such as conventional geotechnical tests including density, moisture content, relative density, seepage, consolidation, direct shear, conventional triaxial tests, true triaxial tests, and centrifuge tests; Field test techniques include field load test, standard penetration test, pressure meter test, vane shear test, static and dynamic penetration test, etc., which are mainly used to solve the problems of the first module. Theoretical calculation is to use mathematical, physical, mechanical equations, complex variable functions and other mathematical and physical principles and methods to establish a calculation method under certain simplified assumptions. It needs to be tested and revised in continuous engineering practice, and is mainly used to solve the problems of the second module. Numerical simulation analysis is to use some existing software, such as ABAQUS, ANSYS, FLAC and PLAXIS, to simulate and analyze the mechanical behavior of soil and structure under various actions, which is mainly used to solve the problem of the third module. The three research approaches and methods should be closely integrated, complement and verify each other, and conduct systematic and in-depth research on a certain soil mechanics problem. However, due to the complexity of soil mechanics problems, experimental conditions, and constraints on scientific research investment, current research in the field of testing is mainly based on existing testing equipment and mature testing techniques, providing reasonable technical indicators for engineering design. In recent years, the research and development of experimental equipment and experimental design for proprietary problems have been developing. The research on soil mechanics should further integrate theoretical calculations, indoor tests, on-site in-situ testing and measurement, model tests, etc., verify and complement each other, draw on each other′s strengths, and carry out long-term systematic exploration through multiple means, methods, and perspectives.

括室内模型试验、现场试验和原位测试量测，如常规土工试验，包括密度、含水率、相对密度、渗流、固结、直接剪切和常规三轴试验、真三轴试验、离心机试验；现场测试技术包括现场载荷试验、标准贯入试验、旁压试验、十字板剪切试验、静（动）力触探等，主要用于解决第一模块问题。理论计算是运用数学、物理、力学方程、复变函数等数学、物理原理和方法，建立一定简化假设条件下的计算方法，需要在不断的工程实践中检验和修正，主要用于解决第二模块问题。数值模拟分析是利用现有的一些软件，如 ABAQUS、ANSYS、FLAC 及 PLAXIS 等模拟分析土体及结构各种作用下的力学行为，主要用于解决第三模块的问题。三种研究途径和方法应紧密结合、互为补充、互为验证，对某一土力学问题开展系统而深入的研究。但由于土力学问题的复杂性、试验条件和科研投入的制约，现阶段试验方面的研究工作主要基于现有试验设备和较为成熟的试验技术，为工程设计提供合理的技术指标。近年来，针对专有问题的试验设备研发和试验设计正在发展。土力学的研究，应进一步将理论计算、室内试验、现场原位测试和量测、模型试验等紧密结合起来，互相验证，互相补充，各取所长，开展多手段、多方法、多角度的长期系统探索。

1.4 Development of Soil Mechanics

The study of soil mechanics originated from people solving engineering practice problems. The development of soil mechanics has roughly gone through four stages:

(1) Origin

From the moment humans emerged, they faced the task of transforming the world and built many magnificent projects, such as China's Great Wall, the Beijing-Hangzhou Grand Canal, the Zhaozhou Bridge, and the Jin Temple, as well as foreign Pyramids, the Taj Mahal, and the Acropolis of Athens, reflecting the rich engineering experience of ancient laboring people. But now it seems that behind it lies a profound theory of soil mechanics.

Driven by the industrial revolution in the 18th century, social productivity developed rapidly, and people's demand for transforming the world increased. The construction of large buildings, bridges, railways, and highways prompted people to study a series of engineering and technical issues related to foundations and roadbeds. In 1773, French scientist Coulomb proposed formulas for the shear strength of soil and the calculation of soil pressure; In 1856, French scholars established the law of soil permeability; In 1885, British scholar Rankine published the theory of soil pressure calculation; French scholar Boussinesq derived the expressions for stress and deformation in soil under the action of vertical concentrated forces on the surface of an elastic half space. The emergence of these classical theories propelled the development of soil mechanics.

(2) Continuation

After World War I, the level of socio-economic development further improved, and people's demand and desire for infrastructure construction further increased, leading to rapid development in soil mechanics research. In 1915, Sweden's Peterson proposed the further development of the circular sliding method for soil

1.4 土力学的发展

土力学的研究源于人们解决工程实践问题。土力学的发展大致经历 4 个阶段：

（1）起源

人类从出现的那一刻起，就面临着改造世界的任务，修建了许多宏伟工程，例如我国的长城、京杭大运河、赵州桥和晋祠等，国外的金字塔、泰姬陵、雅典卫城等，体现了古代劳动人民丰富的工程经验。但现在看来，其背后隐藏的是深厚的土力学理论。

在 18 世纪产业革命的推动下，社会生产力得到快速发展，人民对改造世界的需求增加，大型建筑、桥梁、铁路、公路的兴建，促使人们对地基和路基的一系列工程技术问题进行研究。1773 年，法国科学家库仑提出了土的抗剪强度公式和土压力计算公式；1856 年，法国学者建立了土的渗透定律；1885 年，英国学者朗肯发表了土压力计算理论；法国学者布辛奈斯克推导了弹性半空间表面竖向集中力作用时土中应力和变形的表达式。这些古典理论的出现推动了土力学的发展。

（2）延续

第一次世界大战后，社会经济发展水平进一步提升，人们对基础设施建设的需求和渴望进一步增强，土力学的研究得到迅速发展。1915 年，瑞典的彼得森进一步发展了由费兰纽斯和泰勒

slope stability analysis by Ferranius and Taylor; In 1920, French scholar Prandtl published the sliding surface shape and ultimate bearing capacity formula for foundation shear failure; In 1923, Austrian American Terzaghi discovered the principle of effective stress and proposed the one-dimensional consolidation theory in 1925. In the same year, the first monograph " Soil Mechanics" was published, marking the emergence of soil mechanics as an independent discipline. In 1936, Rendulik discovered the shear dilation, stress-strain nonlinear relationship, and work hardening and softening properties of soil. In 1943, Terzaghi derived the formula for ultimate bearing capacity based on the assumption of sliding surface. The basic framework of soil mechanics has been formed and perfected in this stage.

土坡稳定分析的圆弧滑动法；1920年法国学者普朗特尔发表了地基剪切破坏的滑动面形状和极限承载力公式；1923年，美籍奥地利人太沙基发现了有效应力原理，并于1925年提出了一维固结理论。同年，出版了第一本《土力学》专著，标志着土力学成为一门独立学科。1936年，伦杜利克发现土的剪胀性、应力-应变非线性关系以及土具有加工硬化与软化性质。1943年，太沙基根据滑动面假定推导出了极限承载力公式。这一阶段土力学的基本架构已经形成并得到了完善。

(3) Prosperity

After World War Ⅱ, everything was waiting to be revitalized. A large amount of infrastructure is facing repair and improvement, which has prompted people to invest more enthusiasm in the research of soil mechanics, and the development of soil mechanics has entered a prosperous period. During this period, high-end testing and measurement equipment such as triaxial apparatus, true triaxial apparatus, unsaturated triaxial apparatus, centrifuge, and fiber optic sensors emerged, which led to more efforts beinginvested in the study of soil engineering properties, especially the constitutive relationship of soil. In 1963, Roscoe and others from the University of Cambridge proposed the concept of state boundary surface and established the famous Cambridge elastoplastic model based on it, marking a qualitative leap in people′s understanding of the engineering properties of soil and beginning to describe the mechanical properties of soil from mathematical and physical perspectives. Duncan et al. proposed a hyperbolic nonlinear elastic model; Nonlinear elastic models include *K-G* models and their cross models, high-order nonlinear models (Cauchy, Green, and sub elastic models);

(3) 繁荣

二战结束后，百废待兴。大量基础设施面临修复和完善，促使人们对土力学研究投入更多热情，土力学的发展进入繁荣时期。这一时期，出现了三轴仪、真三轴仪、非饱和三轴仪、离心机、光纤传感器等高端试验和量测设备，使得人们对土的工程性质特别是土的本构关系的研究投入更多的精力。1963年，剑桥大学的罗斯科等人提出了状态边界面概念并据此创立了著名的剑桥弹塑性模型，标志着人们对土的工程性质的认识有了质的飞跃，开始从数学、物理的角度描述土的力学性质。邓肯等人提出了双曲线非线性弹性模型；非线性弹性模型还有*K-G*模型及其交叉模型、高阶非线性模型(Cauchy、Green及次弹性模型)；莱特等人建立了能够反映砂土剪胀性的莱特-邓肯弹塑性

Wright et al. established a Wright Duncan elastoplastic model that can reflect the dilatancy of sandy soil; The Tsinghua University research group led by professor Huang Wenxi proposed the Tsinghua elastoplastic model; Professor Shen Zhujiang proposed an elastoplastic damage model for structural clay. Many scholars have modified these mathematical models by reflecting certain properties of soil based on these classic models, in order to obtain models that are closer to engineering reality. The main development of soil mechanics during this period focused on the study of the constitutive relationship of soil.

模型；以黄文熙教授为首的清华大学研究组提出了清华弹塑性模型；沈珠江教授针对结构性黏土提出了一个弹塑性损伤模型。许多学者在这些经典模型的基础上，反映土的某种性质对这些数学模型进行修正，从而得出更加接近工程实际的模型。这一时期土力学的主要发展集中在土的本构关系研究方面。

(4) Emerging

The development of soil mechanics has gone through origin, continuation, and prosperity, and has formed a relatively systematic and complete theoretical system. But soil mechanics is a highly practical discipline. With the development of the economy and the advancement of technology, a large number of "high", "large", and "superior" engineering constructions have emerged, posing new problems and challenges for soil mechanics research. For example, the design of foundation under super load such as Dubai Tower and Shanghai Tower, the stability and durability of cross sea projects and offshore structures such as Hong Kong-Zhuhai-Macao Bridge and Hangzhou Bay Bridge in complex marine environment, the long-term settlement and control of foundation induced by reciprocating cyclic loads such as high-speed railway and rail transit, and the physical and mechanical properties and bearing capacity of lunar soil in the lunar exploration project in weightlessness environment. The emergence of these new problems has injected vitality into soil mechanics research, promoting the rapid development of soil mechanics in theoretical calculations, numerical simulations, experimental instrument research and development, testing technology, and other fields. Soil mechanics research has entered a new stage of development.

(4) 新兴

土力学的发展经历了起源、延续和繁荣后，已形成了较为系统、完整的理论体系。但土力学是一门实践性很强的学科，随着经济的发展和科技的进步，涌现出一大批“高”“大”“上”的工程建设，为土力学研究提出了新问题、新挑战。例如，迪拜塔、上海中心大厦等超大荷载作用的地基基础的设计，港珠澳大桥、杭州湾大桥等跨海工程及海上结构在复杂海洋环境下的稳定性和耐久性，高速铁路、轨道交通等往复循环荷载诱发基础长期沉降及控制，探月工程中月壤在失重环境下的物理力学性质及其承载力等问题。这些新问题的出现给土力学研究注入了活力，推动着土力学在理论计算、数值模拟、试验仪器研发、测试技术等领域快速发展，土力学研究进入了一个崭新的发展阶段。

Chapter 2 Composition and Physical Properties of Soil

2.1 Introduction

In general, soils are loose accumulation formed by weathering, transportation and sedimentation of rocks on the surface of the earth. Its physical and mechanical properties are greatly influenced by the factors of soil formation. As soils form, substances are being continually removed from and added to the soil with time. Therefore, soils are nature-history production and their profiles vary from place to place.

The formation process of soil is very complex. The rocks on the surface of the earth′s crust undergo weathering under the influence of factors such as sunlight, atmosphere, water, and organisms, causing the rocks to disintegrate and break. Through the dynamic transport of water, wind, glacier, etc., they deposit in various natural environments and form soil. Therefore, it is generally said that soil is a product of rock weathering.

There are three types of weathering in terms of soil formation, which are physical, chemical and biological weathering respectively.

(1) Physical weathering may be caused by the expansion and contraction of rocks from continuous gain and loss of heat, which results in ultimate disintegration. Frequently, water seeps into the pores and existing cracks in rocks. As the temperature drops, the water freezes and expands. The pressure exerted by ice because of volume expansion is strong enough to break down even large rocks. Other physical agents that help disintegrate rocks are glacier ice, wind, the running water of streams, rivers, and ocean waves. This process is called erosion, which refers to the wearing a way of rock by water, wind and ice. It is important to realize that physical weathering belongs to quantitative

第2章 土的组成及其物理性质

2.1 概述

一般来说，土是岩石在地球表面风化、搬运和沉积形成的松散堆积物。土的形成因素对物理力学性质影响很大，随着土的形成，物质会随着时间的推移不断从土中转移或加入其中。因此，土是自然历史的产物，土层剖面因地而异。

土的形成过程非常复杂。地壳表面的岩石在阳光、大气、水和生物等因素的影响下经历风化作用，导致岩石破碎。通过水、风、冰川等的动态运输，它们在各种自然环境中沉积并形成土。因此，人们普遍认为土是岩石风化的产物。

从土的形成角度来看，风化有三种类型，分别是物理风化、化学风化和生物风化。

(1) 物理风化可能是由于岩石因不断获得和损失热量而膨胀和收缩，导致最终解体。水经常渗入岩石中的孔隙和现有裂缝。随着温度的下降，水会冻结并膨胀。由于冰的体积膨胀而施加的压力足以使大块的岩石破碎。其他有助于岩石分解的物理因素有冰川、风、溪流、河流和海浪。岩石被水、风和冰磨损的过程被称为侵蚀。要认识到物理风化属于量变，会产生原始矿物。大块岩石被破碎成小块，化学成分没

change, original minerals will be produced. Large rocks are broken into smaller pieces without any change in the chemical composition. The resulting mineral composition is the same as that of the parent rock, it cracks into cohesionless soil.

有任何变化。由此产生的矿物成分与母岩相同，它会破裂形成无黏性的土。

(2) In chemical weathering, the original rock minerals are transformed into new minerals by chemical reaction. Water and carbon dioxide from the atmosphere form carbonic acid, which reacts with the existing rock mineral to form new minerals and soluble salts. Soluble salts present in the groundwater and organic acids formed from decayed organic matter also cause chemical weathering. It is important to realize that chemical weathering belongs to qualitative change, secondary minerals will be produced. The resulting mineral composition is different from that of parent rock, it cracks into clayed soil.

(2) 在化学风化过程中，原始岩石矿物通过化学反应转化为新矿物。大气中的水和二氧化碳形成碳酸，碳酸与现有的岩石矿物反应形成新的矿物和可溶性盐。地下水中存在的可溶性盐和腐烂有机物形成的有机酸也会引起化学风化。要认识到化学风化属于质变，会产生次生矿物。由此产生的矿物成分与母岩不同，会碎裂形成黏性土。

(3) Biological weathering refers to weathering caused by organisms like animals, plants, fungi and bacteria. While certain forms of biological weathering, such as the breaking of rock by tree roots, are sometimes categorized as either physical or chemical. Biological weathering can work hand in hand with physical weathering by weakening rock or exposing it to the forces of physical or chemical weathering. Biological weathering leads to the formation of organic matter and changes the structure of the soil.

(3) 生物风化是指由动物、植物、真菌和细菌等生物引起的风化。而某些形式的生物风化，如树根破坏岩石，有时被归类为物理或化学风化。生物风化可以通过削弱岩石或使其暴露在物理或化学风化作用下，协同作用。生物风化会形成有机物并改变土的结构。

Transportation and sedimentation are the next processes of soil formation. The products of weathering may stay in the same place or may be moved to other places.

运输和沉积是土形成的下一个过程。风化产物可能留在原地，也可能被转移到其他地方。

The soils formed by the weathered products at their place of origin are called residual soils. An important characteristic of residual soil is the gradation of particle size. Fine-grained soil is found at the surface, and the grain size increases with depth. At greater depths, angular rock fragments may also be found. They have rough surface, much pointedness, thickness odds and no lamination.

风化产物在其来源地形成的土称为残积土。残积土的一个重要特征是粒径的级配。表层为细粒土，粒径随深度增加而增大。在更深的地方，也可能发现有棱角的岩石碎片，它们表面粗糙，尖端多，厚度不均匀，没有层压。

If the products of weathering are moved to other places by ice, water, wind and gravity, the soils are called transported soils.

如果风化产物被冰、水、风和重力移动到其他地方，这些土就被称为运积土。

Soil formed through weathering, which accepts the natural power action to transport to different area as deposit. The transported soils may be classified into several groups, depending on their mode of transportation and deposition. Such as:

风化形成的土，在自然力作用下作为沉积物搬运到不同地区。根据运输和沉积方式，搬运的土可分为几组。例如：

Glacial soils formed by transportation and deposition of glaciers. Its particle size varies greatly and the soil is uneven.

冰积土——由冰川的运输和沉积形成，其颗粒粗细变化大，土质不均匀。

Alluvial soils transported by running water and deposited along streams. Due to their transportation over long distances, the particles often exhibit a high degree of rounding as well as distinct lamination.

冲积土——由流水运输并沿溪流沉积形成，由于长距离的搬运，颗粒通常高度接近球形并有明显的分层。

Lacustrine soils formed by deposition in quiet lakes. This type of soil is often accompanied by organic matter formed through biochemical processes.

湖积土——在静止的湖泊中沉积形成，这类土常伴有生化作用所形成的有机物。

Marine soils formed by deposition in the seas.

海积土——在海洋中沉积形成。

Aeolian soils transported and deposited by wind.

风积土——由风力运输并沉积形成。

Colluvial soils formed by movement of soil from its original place by gravity, such as during landslides.

崩积土——在重力作用下，土从原处移动而形成，比如滑坡。

Soil is nature-history production, which is formed by rock in different conditions after weathering. It has natural variability. The particles of the soil mass have no cohesion or have a certain degree of cohesion, with a large number of pores, exhibiting granular characteristics. As the mixture of solid particles, water and gas, soils can be tri-phase composition.

土是岩石风化后在不同条件下形成的自然历史产物，具有自然变异性。土体颗粒无黏聚力或具有一定程度的黏聚力，具有大量的孔隙，表现出颗粒状特征。土是固体颗粒、水和气体的混合物，即三相组成。

2.2 Tri-phase Components of Soil

2.2 土的三相组成

The composition of natural soil includes solid phase, liquid phase, and gas phase as shown in Figure 2.1. A given volume of soil in natural occurrence consists of solid particles and the void spaces between the particles. The void space may be filled with air and/or

天然土的组成包括固相、液相和气相，如图 2.1 所示。自然状态下给定体积的土由固体颗粒和颗粒之间的孔隙组成。孔隙可以被空气或水填充，因此，土是

water, hence, soil is a tri-phase system. If there is no water in the void space, it is dry soil. If the entire void space is filled with water, it is referred to as a saturated soil. However, if the void is partially filled with water, it is moist soil.

一个三相系统。如果孔隙中没有水，那就是干燥的土。如果孔隙部分被水充满，就是湿润的土。如果整个孔隙空间都充满了水，则称为饱和土。

Solid phase—solid particles compose the soil framework. They are the primary components of soil, providing structural support and influencing various soil properties. They have final effect to the soil's characteristic.

固相——固体颗粒组成了土骨架。它们是土的主要成分，提供结构支撑并影响各种土的性质。它们对土的特性有决定性影响。

Voids between particles, where the voids are filled with water, which has significant effect to soil. And gas phase is in the rest void of soil skeleton, has secondary effect to soil.

土的颗粒之间的孔隙中充满了水，对土的性质有重大影响。而气相存在于土骨架的其余孔隙中，对土的性质有次要影响。

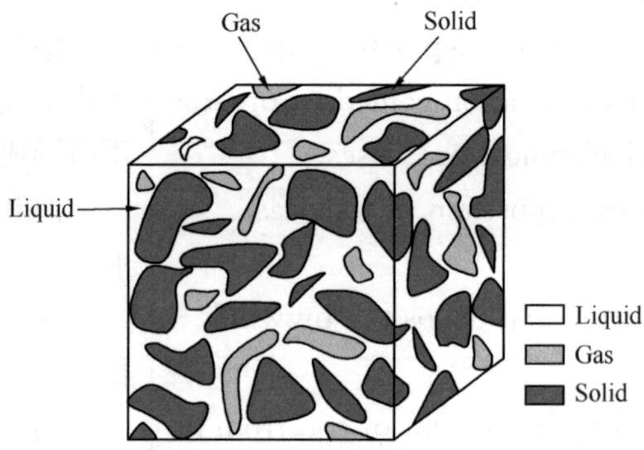

Figure 2.1 The composition of soil

2.2.1 Solid Phase

The solid particles of soil are the solid phase of soil. The size, shape, mineral components, and grading of soil particles have a significant impact on the physical and mechanical properties of soil.

2.2.1 固相

土的固体颗粒是土的固相。土颗粒的大小、形状、矿物成分和级配对土的物理力学性质有重要影响。

1. Size grading

Size grading refers to different particle sizes contained in the soil.

1. 颗粒级配

颗粒级配是指土中含有的不同粒径。

The mineral components form soil particles and vary with the mineral components of parent rock and weathering. Generally speaking, size grading and shapes affect the physical state and mechanical characteristics of soil.

矿物成分形成土颗粒，并随着母岩和风化的矿物成分而变化。一般来说，颗粒级配和形状会影响土的物理状态和力学特性。

The size of solid particles is related to the soil-

固体颗粒的大小与土形成矿

forming minerals and it reflects the property of soil. The size of soil particles is called grain size, expressed in terms of particle size. The soil particles within a certain particle size range are called fraction. Each fraction exhibits certain qualitative changes with different boundary sizes. The boundary size for dividing fraction is called the boundary particle size. We classify the particles with the similar size and property as grain group.

物有关，反映了土的性质。土粒的大小称为粒度，以粒径表示。介于一定粒度范围内的土粒，称为粒组。各个粒组因分界尺寸的不同，而呈现出一定质的变化。划分粒组的分界尺寸称为界限粒径。我们将尺寸和性质相似的颗粒归类为同一粒组。

For instance, following "Standard of Engieering Classification of Soil" GB/T 50145—2007, we can divide soils into two groups. If particle size is smaller than or equal to 0.075mm it is known as fine soil (silt or clay). Soil particles with a particle size of 60 mm or smaller are referred to as coarse soil, while those with a particle size larger than 60 mm are termed as very coarse soil.

例如，根据《土的工程分类标准》GB/T 50145—2007，我们可以将土分为两类。如果粒径小于等于 0.075mm，则称为细粒土（粉土或黏土）。粒径小于等于 60mm 称为粗粒土，粒径大于 60mm 称为巨粒土。

The particle size distribution of a coarse soil can be determined by the method of sieving. The soil sample is passed through a series of standard test sieves having successively smaller mesh sizes. The mass of soil retained in each sieve is determined, and the cumulative percentage by mass passing each sieve is calculated.

粗粒土的粒度分布可以通过筛分法确定。使用一系列标准试验筛筛土样，筛孔尺寸从上到下依次减小。称量每个筛网中保留的土质量，并计算通过每个筛网的土的累积质量百分比。

The particle size distribution of a fine soil can be determined by the method of sedimentation. This method based on the relationship between the settling velocity and particle size of spherical particles in suspension: the larger the particle, the greater the settling velocity, and vice versa, which is known as Stoke′s law.

细粒土的粒度分布可以通过沉降法来确定。这种方法与球形颗粒在悬浮液中的沉降速度与粒径存在一定关系：颗粒越大，沉降速度越大，反之亦然，这就是斯托克斯定律。

2. Particle size distribution curve

2. 颗粒级配曲线

To analyze the particle size distribution composition of soil samples, let′s discuss one example. A sample of dry sand with weight of 100g is fully shaken on a series of standard test sieves.

为分析土样的颗粒级配组成，讨论一个例子。将质量为 100g 的干砂样品在一系列标准试验筛上充分摇动。

The soil is shaken through a stack of sieves with openings of decreasing size from top to bottom, a pan is placed below the stack. After the soil is shaken, the mass of soil retained on each sieve is determined. Base on the mass of the dry soil retained on each sieve, we

土样通过一系列筛孔从上到下逐渐变小的筛子，摇动筛子，在筛子下面放置一个盘子。土筛分完成后，确定每个筛子上保留的土质量。根据每个筛子上残留

are going to do the sieve analysis.

的干土质量，进行筛分分析。

Firstly, determine the mass of the soil retained on each sieve corresponding to different sieve opening size and the mass of the soil in the pan, which are shown in Table 2.1. Secondly, determine mass of each particle group.

首先，确定不同筛孔尺寸对应的每个筛子上保留的土质量和盘中的土质量，如表 2.1 所示。其次，确定每个粒组的质量。

Next, determine percentage of soil passing the sieve (%), cumulative mass of soil passing each sieve (%) is in the last column.

接下来，确定通过筛子的土百分比（%），通过每个筛子的土累积质量（%）（表 2.1 最后一列）。

Therefore, to illustrate how to figure it out, here is one example. For instance, the weight of soil retained on 5mm is 3g, the weight of soil retained on 2mm is 7g, the mass percentage of soil comprising with particles smaller than 2mm is equal to [100－(7＋3)]%＝90/100＝90%, with the same way, all results are calculated as shown in the last column of Table 2.1.

为了说明如何计算，这里举个例子。若在 5mm 筛子上保留的土质量为 3g，2mm 筛子上保留的土质量为 7g，含有小于 2mm 颗粒的土质量百分比等于 [100－(7＋3)]%＝90%，所有计算结果见表 2.1 最后一列。

Particle size distribution **Table 2.1**

Sieve opening (mm)	Mass of soil retained on each sieve (g)	Mass of every particle group (%)	Cumulative mass of soil passing each sieve (%)
10.0	0	0	100
5.0	3	3	97
2.0	7	7	90
1.0	10	10	80
0.5	14	14	66
0.25	16	16	50
0.075	22	22	28
0.05	6	6	22
0.025	7	7	15
0.01	7	7	8
Bottom head	8	8	0

For diameter of soil is smaller than 0.075mm, sedimentation analysis is used. Sedimentation analysis is

对于直径小于 0.075mm 的土，采用沉降分析。沉降分析基

based on the principle of sedimentation of soil grain in water. When a soil sample is dispersed in water, the particles settle at different velocities. It is assumed that all the soil particles are spheres and that the velocity of soil particles can be expressed the proportional function of the diameter of soil particles.

于土颗粒在水中的沉降原理。当土样分散在水中时，颗粒以不同的速度沉降。假设所有土颗粒都是球体，土颗粒的速度可以表示为土粒直径的比例函数。

In this example the results are shown in Table 2.1. Now we use the results to plot the curve shown in Figure 2.2. In the curve, the vertical coordinate represents the percentage by mass of particles smaller than the size and the horizontal coordinate represents particle size with a semi-logarithmic plot which is satisfied with the great difference between particle size. This Figure 2.2 is referred to as the particle size distribution curve.

该例结果如表 2.1 所示。绘制如图 2.2 所示的曲线。曲线纵坐标表示小于粒径的颗粒的质量百分比，横坐标表示粒径，其半对数满足粒径之间的差异巨大。图 2.2 被称为颗粒级配曲线。

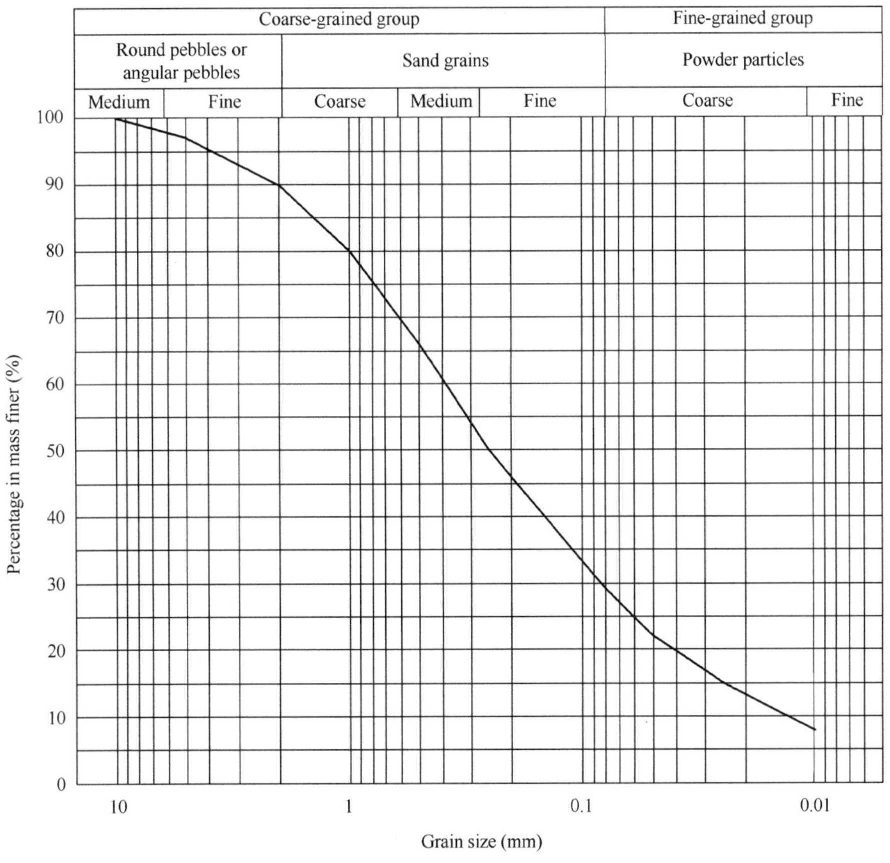

Figure 2.2 Particle size distribution curve

In this curve d_{10}, d_{30}, and d_{60}: denote 10%, 30% and 60% of the particles which is smaller than that size

在这条曲线中，d_{10}、d_{30} 和 d_{60} 表示小于某一粒径的质量百

in grading curve. Where d_{10}: effective grain size; d_{30}: median grain size; d_{60}: constrained grain size.

分比为 10%、30% 和 60%时对应的粒径，分别称为有效粒径、中值粒径和限制粒径。

Relevant indexes for grading assessment can be get from particle size distribution curve. As shown in Figure 2. 2, low degree of slope, represents the soil containing a wide range of particle size. We use C_u to express non-uniform coefficient.

从颗粒级配曲线可以得到级配评价的相关指标。如图 2. 2 所示，坡度小表示土含有粒径范围更宽。用 C_u来表示不均匀系数。

Generally speaking, the larger the non-uniformity coefficient, the higher degree of density can be obtained through compacting. In engineering practice, we use non-uniform coefficient C_u to demonstrate the degree of irregularity.

一般来说，不均匀系数越大，通过压实可以获得更高的密度。在工程实践中，使用不均匀系数 C_u来表示不规则程度。

$$C_u = \frac{d_{60}}{d_{10}} \tag{2.1}$$

$C_u \geqslant 5$, nonuniform soil;
$C_u < 5$, uniform soil.

$C_u \geqslant 5$，不均匀土；
$C_u < 5$，均匀土。

But some well graded soil, some particles with middle size are absent and the flowing water can carry small particles through the pores. In this case, the shape of grading curves can be represented with curvature coefficient C_c.

但一些级配良好的土中不存在中等大小的颗粒，流动的水可以携带小颗粒通过孔隙。在这种情况下，级配曲线的形状可以用曲率系数 C_c表示。

$$C_c = \frac{d_{30}^2}{d_{60} d_{10}} \tag{2.2}$$

$C_c = 1 \sim 3$, continuous soil;
$C_c > 3$ or $C_c < 1$, discontinuous soil.

$C_c = 1 \sim 3$，级配连续的土；
$C_c > 3$ 或 $C_c < 1$，级配不连续的土。

The summary is as follows:

1) The content of each fraction. It is used to classify coarse-grained soil and assess engineering property of soil.

2) The characteristic particle sizes. It is used for the selection of building material and assessment of particle grading.

3) Non-uniform coefficient C_u is used to decide soil's degree of irregularity: $C_u \geqslant 5$, nonuniform soil; $C_u < 5$,

总结如下：

1）每个粒组的含量。用于对粗粒土进行分类，并评估土的工程性质。

2）特征粒径。用于进行建筑材料的选择和颗粒级配的评估。

3）不均匀系数 C_u用于确定土的不规则程度：$C_u \geqslant 5$，不均

uniform soil.

4) Curvature coefficient C_c is used to decide soil's continuous degree: $C_c = 1 \sim 3$, grain group continuous soil; $C_c > 3$ or $C_c < 1$, grain group discontinuous soil.

5) Non-uniformity coefficient C_u and curvature coefficient C_c is used to decide soil's graded quality:

if $C_u \geqslant 5$ and $C_c = 1 \sim 3$, good graduation soil;

if $C_u < 5$ or $C_c > 3$ or $C_c < 1$, bad graduation soil.

As mentioned before, the mineral components which form soil particles vary with the vary of mineral components of parent rock and weathering. Minerals include two major types: one type is original mineral, which includes quartz, feldspar. Another type is secondary mineral, which is formed through chemical weathering of original mineral, and has the components different from parent rock. Secondary mineral consists mainly of clay mineral.

As a natural production, soil consists of countless particles which have different shapes. The shape of particles present in a soil mass is equally as important as the particle size distribution because it has significant influence on the physical properties of a given soil. However, not much attention is paid to particle shape because it is more difficult to measure. Generally, the majority shapes for original mineral are circle, for secondary mineral, needle slice flat shape can be seen from scanning electron microscopy.

2.2.2 Liquid Phase

Then we'll talk about water in soil which is the liquid phase, and its types and quantities have important influence upon the state and porosity of soil. Water in soil can occur in the form of combined water and in the form of free water. Combined water: water absorbed on soil grain surface; Free water: water out of electric field gravitation.

匀土；$C_u<5$，均匀土。

4）曲率系数 C_c 用于确定土的连续程度：$C_c=1\sim3$，颗粒连续土；$C_c>3$ 或 $C_c<1$，颗粒不连续土。

5）不均匀系数 C_u 和曲率系数 C_c 用于确定土的级配质量：

当 $C_u\geqslant5$，$C_c=1\sim3$ 时，土级配良好；

如果 $C_u<5$ 或 $C_c>3$ 或 $C_c<1$，则土级配较差。

如前所述，形成土颗粒的矿物成分随着母岩和风化的矿物成分变化而变化。矿物包括两种主要类型：一种是原始矿物，包括石英、长石。另一种是次生矿物，是通过原始矿物的化学风化形成的，其成分与母岩不同。次生矿物主要由黏土矿物组成。

作为一种自然产物，土由无数形状各异的颗粒组成。土体中颗粒的形状与粒径分布同样重要，因为它对给定土的物理性质有重大影响。然而，由于颗粒形状更难测量，因此没有受到太多关注。一般来说，原始矿物的大多数形状是圆形的，对于次生矿物，扫描电镜可以看到针片状的扁平形状。

2.2.2 液相

接下来讨论土中的水，它属于液相，其类型和数量对土的状态和孔隙度有重要影响。土中的水可以以结合水和自由水的形式存在。结合水为土颗粒表面吸收的水；自由水为电场引力作用下的水。

The individual soil particles are covered with a very thin surface of combined water by the electrochemical force between soil particles. Fine particles have negative charges, which create electronic field around them and make water molecules polarized, and water molecules include hydrogen atom having positive electricity and oxygen atom having negative electricity which array directionally around soil particles. Cations in water such as Calcium ion（Ca^{2+}）are attracted by the soil particles, so soil particles are covered with a very thin film of water.

通过土颗粒之间的电化学力，单个土颗粒被非常薄的结合水表面覆盖。细颗粒带负电荷，这会使它们周围产生电场，使水分子极化，水分子包括带正电的氢原子和带负电的氧原子，它们在土颗粒周围定向排列。水中的阳离子（如钙离子 Ca^{2+}），被土颗粒吸引，因此土颗粒被一层非常薄的水覆盖。

The combined water can be classified into strong and feeble absorbed water base on the distance from soil particles. Strong absorbed water is known to be bound or attached rigidly to the soil particles with an immerse physical force. Feeble absorbed water is called film water, which is held by molecular forces of considerable intensity. Free water is far from soil particles, and not affected by physical force, and remove from the bulk. It includes capillary water and gravitational water. Capillary water exists between solid and gas. Gravitational water under gravitation and surface tension, can move on soil grain interspace freely. Gravitational water can flow in soil.

根据与土颗粒的距离，结合水可分为强结合水和弱结合水。强结合水会通过浸没的物理力与土颗粒牢固地结合或附着在土颗粒上。弱结合水被称为薄膜水，它由相当大的分子力保持。自由水远离土壤颗粒，不受物理力的影响，从土中去除，包括毛细水和重力水：毛细水存在于固体和气体之间；重力水在重力和表面张力的作用下，可以在土颗粒间隙中自由移动，即重力水可以在土中流动。

2.2.3 Gas Phase

The gas phase occupies that part of the voids which is not occupied by liquid, and it is mostly air and sometimes CO_2, but with lower content of O_2 and N_2. If the gas connected with the air, it's called free gas. It affects the properties of soil little. The gas that exists in the form of bubbles in the voids of soil and is isolated from the atmosphere is called close gas, it increases elasticity of soil, block seepage pathway so as to reduce the permeability and extend the time when soil retains stability with external action.

2.2.3 气相

气相占据了孔隙中未被液体占据的部分，其组成主要为空气，有时也存在 CO_2，但 O_2 和 N_2 含量较少。如果气体与空气相连，称为自由气体，它对土的性质影响较小。在土的孔隙中以气泡形式存在且与大气隔绝的气体被称为封闭气体，它增加了土的弹性性能，阻塞了渗透通道，从而降低渗透性，并延长土在外部作用下保持稳定的时间。

This is all for tri-phases components of soil. We have learned the composition of natural soil may include

这就是土的三相组成。所以，天然土的组成包括构成土骨

solid phase composing soil framework and liquid phase, gas phase which occupy the voids of the soil.

架的固相和占据土孔隙的液相、气相。

2.3 Phase Relationships

Section one and two present the geologic processes by which soil is formed, the description of the limits on the sizes of soil particles, and the mechanical analysis of soils. Hence it is important in all geotechnical engineering works to establish relationships between mass and volume in a given soil mass.

In this section we will discuss the following: First to define and develop mass relationships such as mass density, specific gravity and water content of soil in combination with the volume relationships. These indexes can be obtained by experiment. We call them Basic Physical Indexes. Next to define and develop dimensionless volume relationships such as void ratio, porosity, and degree of saturation. These indexes are named calculated or indirect indexes.

2.3.1 Basic Physical Indexes

Figure 2.3 shows an element of soil of volume V and mass m as it would exist in a natural state. To develop the mass-volume relationships, we must separate the three phases—that is, solid, water, and air.

2.3 相关系

第 1 节和第 2 节介绍了土形成的地质过程、土颗粒界限尺寸的定义以及土的力学分析。在岩土工程中，给定土体中质量和体积之间的关系非常重要。

本节讨论以下内容：首先，结合体积关系，定义和延伸质量关系，如土的质量密度、相对密度和含水量。这些指标可以通过试验获得，称为基本物理指标。接下来，定义和延伸无量纲体积关系，如孔隙比、孔隙度和饱和度，这些指标被称为计算指标或间接指标。

2.3.1 基本物理指标

图 2.3 为自然状态下体积为 V、质量为 m 的一个土体单元。为了建立质量-体积关系，必须将三相分离，即固体、水和空气。

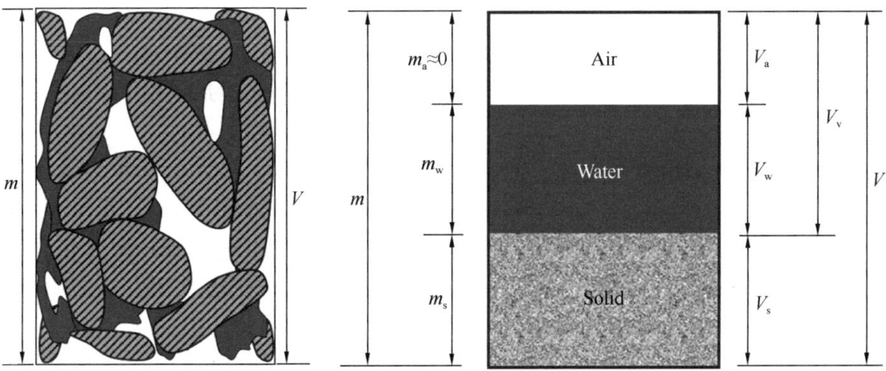

Figure 2.3 Soil composition

Thus, the total volume of a given soil sample shown in right-hand side can be expressed as:

因此，右侧图中所示给定土样的总体积可以表达为式 (2.3)。

$$V = V_s + V_v = V_s + V_w + V_a \tag{2.3}$$

where V_s is volume of soil solids;
V_v is volume of voids;
V_w is volume of water in the voids;
V_a is volume of air in the voids.

式中，V_s为土的固体体积，V_v为孔隙体积，V_w为孔隙中水的体积，V_a为孔隙中气体的体积。

The total mass of soil in left-hand side, assuming that the mass of the air is negligible, we can express the total mass of the sample as:

假设空气的质量可以忽略不计，左侧土样的总质量表示为式(2.4)。

$$m = m_s + m_w \tag{2.4}$$

where m_s is mass of soil solids, m_w is mass of water.

式中，m_s为土的固体质量，m_w为水的质量。

The components of soil can be represented by a phase diagram as shown in Figure 2.3, from which the following relationships are defined. The most useful relationships between phase masses are density, specific gravity and water content. These indexes must be obtained by tests, which are named direct indexes.

土的成分可以用图 2.3 所示的相图来表示，从中可以定义以下关系。相质量之间最有用的关系是密度、相对密度和含水量。这些指标必须通过试验获得，称为直接指标。

Bulk density is the ratio of the total mass to the total volume. Its symbol is ρ. It can be expressed as follows:

密度是总质量与总体积之比，用符号 ρ 表示，可以表达为式（2.5）。

$$\rho = \frac{m}{V} = \frac{m_s + m_w}{V_s + V_w + V_a} \tag{2.5}$$

The unit of bulk density is kg/m^3 or g/cm^3.

其单位为 kg/m^{3}或 g/cm^3。

The relative index of bulk density is the unit weight of soil which is the ratio of the total weight to the total volume. The symbol is γ, it can be expressed by:

重度是指单位体积的重量，即总重量与总体积之比，符号为 γ，可表达为式（2.6）。

$$\gamma = \rho g \tag{2.6}$$

Its unit is kN/m^3; g is the acceleration of gravity.

单位为 kN/m^3；g 为重力加速度。

The second basic index is specific gravity of soil particles, we use G_s to denote specific gravity of soil particles. The specific gravity of soil particles is the ratio of the density of soil particles to the density of pure distilled water at four degrees Celsius.

第二个基本物理指标是土粒相对密度，G_s表示土粒相对密度。土粒相对密度是质量与 4℃纯蒸馏水质量的比值，用式(2.7) 表示。

$$G_s = \frac{m_s}{V_s \cdot \rho_w^{4℃}} = \frac{\rho_s}{\rho_w^{4℃}} \tag{2.7}$$

Specific gravity is a dimensionless parameter used to describe the density of soil particles. In general, the particle specific gravity range of different soils is different. For cohesive soil, the specific gravity range is approximately 2.70

土粒相对密度是一个无量纲参数，用来描述土粒的密度。一般来说，不同土体的相对密度范围是不同的。对于黏性土来说，

~2.75; for sandy soil, the specific gravity is approximately 2.65. Due to the density of water at 4℃ being 1g/cm^3. The specific gravity of soil particles is numerically equal to the density of soil particles.

相对密度范围大约是 2.70~2.75；砂土的相对密度大约为 2.65。由于水在 4℃时的密度为 1g/cm^3，所以土粒相对密度数值上等于土粒的质量。

The third basic index is water content, also called moisture content, the water content is the ratio of the mass of water to the mass of solids in the soil. It can be expressed by percent as follows:

第三个基本物理指标是土的含水量，也称为土的含水率。含水量是土中水质量与固体质量的比值，土的含水量用百分比可表示为式（2.8）。

$$w = \frac{m_w}{m_s} \times 100\% \tag{2.8}$$

The water content is also a dimensionless parameter, soil has a large variation range of water content. The amount of water content can reach or exceed 100%.

含水量也是一个无量纲参数，土含水量的变化范围很大。含水量的量值可以达到或超过 100%。

2.3.2 Calculated Physical Indexes

2.3.2 计算指标

Calculated Physical Indexes can be calculated from basic indexes such as void ratio, porosity and degree of saturation.

计算物理指标可以根据孔隙比、孔隙度和饱和度等基本指标进行计算。

Void ratio e

孔隙比 e

Void ratio is the ratio of the volume of voids to the volume of solids:

孔隙比是土中孔隙体积与固体体积之比，可表示为式（2.9）。

$$e = \frac{V_v}{V_s} \tag{2.9}$$

It is an index indicating the content of pores in the soil.

它是表示土中孔隙含量的指标。

Porosity n

孔隙率 n

Porosity is also an index indicating the content of pores in the soil.

孔隙率也是表示土中孔隙含量的指标。

The porosity is the ratio of the volume of voids to the total volume of the soil. It can be expressed by percent as follows:

孔隙率是指孔隙体积与土的总体积之比。可以用百分比表示为式（2.10）。

$$n = \frac{V_v}{V} \times 100\% \tag{2.10}$$

Both void ratio and porosity are indexes used to express the pore volume content, and they are interrelated

孔隙比和孔隙率都是表示孔隙体积含量的指标，它们之间的

as follows:

关系为式（2.11）。

$$n=\frac{e}{1+e},\ e=\frac{n}{1-n} \tag{2.11}$$

Degree of saturation S_r

The degree of saturation (S_r) is defined as the ratio of the volume of water to the volume of voids.

饱和度 S_r

饱和度是水的体积与孔隙体积之比，可表示为式（2.12）。

$$S_r=\frac{V_w}{V_v}\times 100\% \tag{2.12}$$

It indicates the degree to which the pores are filled with water:

For completely dry soil: $S_r=0$; For fully saturated soil: $S_r=100\%$.

Water content and the degree of saturation are indexes that indicate the degree of water in the soil.

Saturated density ρ_{sat}

Saturated density is the density when the soil is fully saturated with water. Saturated density can be denoted by ρ_{sat}. Expressed as follows:

它表示孔隙被水填充的程度：

对于完全干燥的土 $S_r=0$；对于完全饱和的土 $S_r=100\%$。

含水量和饱和度都是表示土中水充满程度的指标。

饱和密度 ρ_{sat}

饱和密度是指土被水完全饱和时的密度。饱和密度可表示为式（2.13）。

$$\rho_{sat}=\frac{m_s+\rho_w V_v}{V} \tag{2.13}$$

Dry density ρ_d

Generally, to solve earthwork problems, one must know the mass per unit volume of soil, excluding water. This mass is referred to as the dry density, ρ_d. Thus, dry density is the density of the soil when it is completely dried. It is given by:

干密度 ρ_d

通常，要解决土方工程问题，必须知道土的单位体积质量（不包括水分），此指标被称为干密度 ρ_d。干密度是土在完全干燥时的密度，用式（2.14）表示。

$$\rho_d=\frac{m_s}{V} \tag{2.14}$$

So far, we know three densities: bulk density, saturated density and dry density. What's their relationship? Apparently, they have the following relationship.

到目前为止，有三种密度：体积密度、饱和密度和干密度。它们之间有什么关系呢？显然，它们之间存在式（2.15）所述的关系。

$$\rho_{sat}\geqslant\rho\geqslant\rho_d \tag{2.15}$$

Density refers to the mass per unit volume, and the unit weight is the ratio of total weight to total volume. In general, density (g/cm^3) can be converted to unit weight (kN/m^3) by multiplying by the acceleration

密度是指单位体积的质量，单位重量是总重量与总体积的比值。一般情况下，密度（g/cm^3）乘以重力加速度 g 可以换

due to gravity (g) . Effective unit weight equals saturated unit weight minus the unit weight of water.

算成重度（kN/m^3）。有效重度等于饱和单位重量减去水的单位重量，表示为式（2.16）。

$$\gamma' = \gamma_{sat} - \gamma_w \tag{2.16}$$

So, the unit weight of soil has the following relationship.

所以土的单位重量的关系为式（2.17）。

$$\gamma_{sat} \geqslant \gamma \geqslant \gamma_d \geqslant \gamma' \tag{2.17}$$

As mentioned before, there are nine indexes used. Generally, if basic indexes are known, the rest of them can be derived by their relationships as follows.

上文共讲述了九个指标。一般来说，如果已知基本指标，则可以根据它们之间的关系推导出其余的指标。

We move m_s to the left-hand side, equation (2.18) can be obtained. We transfer m_w to the left-hand side, equation (2.19) can be given.

将 m_s 放到左边，得到式（2.18），将 m_w 移到左边，得到式（2.19）。

$$G_s = \frac{m_s}{V_s \cdot \rho_w} = \frac{m_s}{p_w} \rightarrow m_s = G_s\rho_w \tag{2.18}$$

$$w = \frac{m_w}{m_s} \times 100\% \rightarrow m_w = wm_s = wG_s\rho_w \tag{2.19}$$

$$e = \frac{V_v}{V_s} \tag{2.20}$$

For $V_s = 1$, void ratio $e = V_v$. As bulk density is the ratio of the total mass to the total volume, it can be expressed as:

在式（2.20）中，如果 $V_s = 1$，孔隙比 $e = V_v$。体积密度为总质量与总体积之比，可表示为式（2.21）。

$$\rho = \frac{m}{V} = \frac{m_s + m_w}{V_s + V_w + V_a} = \frac{m_s + m_w}{1 + V_v} \tag{2.21}$$

Then we substitute equations (2.18)、(2.19) and (2.20) into equation (2.21), the equation (2.22) will be acquired.

将式（2.18）～式（2.20）代入式（2.21），得到式（2.22）。

$$\rho = \frac{G_s\rho_w + wG_s\rho_w}{1 + V_v} \tag{2.22}$$

We transfer e to left-hand side, the expression of calculated index e can be get as following:

将孔隙比 e 移到左边，计算 e 的表达式为式（2.23）。

$$e = \frac{(1 + w)G_s\rho_w}{\rho} - 1 \tag{2.23}$$

It also can be denoted by unit weight γ:

也可以用重度 γ 表示为式（2.24）。

$$e = \frac{(1 + w)G_s\gamma_w}{\gamma} - 1 \tag{2.24}$$

Relationship among dry density, void ratio, and

干密度、孔隙率和土粒相对

specific gravity of soil can be derived in a similar manner. Using the definitions of G_s, w and e, the relationship among dry density, void ratio, and specific gravity of soil can be expressed as following equation (2.25):

密度之间的关系可以用类似的方法推导出来。用 G_s、w、e 的定义，土体的干密度、孔隙率、土粒相对密度之间的关系可表示为式(2.25)。

$$\rho_d = \frac{G_s\rho_w}{1+e} \tag{2.25}$$

2.4 Soil Physical Condition

In this sector, Soil Physical Condition Index will be explained. Physical conditions and fabrics of soil during formation process ultimately determine permeation, deformation and strength characteristics of soil. Natural soil is a very complex physical system. Soil physical properties index is about the relationships between three phase masses. Normally in engineering practice, the stiffness and compaction are needed to be provided. Physical condition index is to solve this problem.

The characteristics of coarse-grained soil depend on the compactness of the soil. For cohesive soil, the consistency of the soil affects its mechanical properties. As is well known, soil exhibits fluidity and viscosity when it has a highwater content, during which time the soil consolidation process requires a shorter duration, often manifesting as instant settlement.

At lower water contents the soil no longer flows, it is stiff, sticky and plastic. As the moisture content is decreased further the soil loses its stickiness and plasticity and becomes crumbly or friable, it feels soft to touch. Finally, in the dry state it is hard and coherent, the sensation of harshness is experienced when the soil is rubbed between the fingers.

Soil physical condition affects mechanical characteristic of soil. Now we will explain them respectively.

2.4.1 Compaction of Coarse-grained Soil

For coarse-grained soil, the void ratio of a state mass of the identical cohesionless spheres depends on the manner in which the spheres are arranged. Natural sands are found with void ratio varying from about 0.25 to 0.5.

2.4 土的物理状态

本节将解释土的物理状态指标。形成过程中土的物理状态和结构最终决定了土的渗透性、变形和强度特性。天然土是一个非常复杂的物理系统。土的物理性质指标是三相质量之间的关系。通常在工程实践中，需要提供土的刚度和密实度。物理状态指标则是为了解决这个问题而提出的。

粗粒土的特性取决于土的密实度。对于黏性土，土的密实度会影响其力学性能。当土的含水量高时，土会表现出流动性和黏性，土的固结过程更快，通常表现为瞬时沉降。

含水量较低时土不再流动，变得坚硬、黏稠和可塑。随着含水量的进一步降低，土失去了黏性和可塑性，变得易碎，摸起来很柔软。在干燥状态下土坚硬而连贯，当用手指揉搓土体时，会感到粗糙。

土的物理状态影响土的力学特性，下面分别进行解释。

2.4.1 粗粒土的密实度

粗粒土的孔隙比取决于无黏聚力的相同球体的排列方式。天然砂土的孔隙比约为 0.25～0.5。

The porosity of a natural sand deposit depends on the shape of the grains, the uniformity of grain size, and the conditions of sedimentation.

天然砂土沉积物的孔隙比取决于颗粒的形状、颗粒大小的均匀性以及沉积条件。

Because of the great influence of the shape of the grains and degree of uniformity, the porosity itself does not indicate whether a soil is loose or dense. For the same of soil, the porosity can be used to compare the density of soil in its loosest and densest states.

由于颗粒形状和均匀度的影响很大，孔隙比本身并不能表明土是松散的还是密实的。对于相同的土，孔隙比可用于比较土在最松散和最密实状态下的密实状态。

The actual void ratio of a soil lies somewhere between the possible minimum and maximum values e_{min} and e_{max}. A convenient measure of the state of compaction is provided by a relationship between the void ratio values which is termed the relative compaction. Relative compaction applies only to sand, not silt or clay. The relative compaction is given as the following expressions:

土的实际孔隙比介于可能的最小值 e_{min} 和最大值 e_{max} 之间。孔隙比之间的关系提供了一种方便的密实状态测量方法，称为相对密实度。相对密实度仅适用于砂土，而不适用于粉土或黏土，可表示为式（2.26）。

$$D_r = \frac{e_{max} - e}{e_{max} - e_{min}} \tag{2.26}$$

Where e is void ratio of soil in the normal state;

e_{max} is void ratio of soil in the loosest state;

e_{min} is void ratio of soil in the densest state.

式中，e 为土在正常状态下的孔隙比，e_{max} 为土在最松散状态下的孔隙比，e_{min} 为土在最密实状态下的孔隙比。

The most effective method of identifying cohesionless soils is the soil vibration method. It can be expressed by density as shown in:

识别无黏性土的最有效方法是振动法，可以用密度表示为式（2.27）。

$$D_r = \frac{(\rho_d - \rho_{dmin})\rho_{dmax}}{(\rho_{dmax} - \rho_{dmin})\rho_d} \tag{2.27}$$

where ρ_d is dry density in the normal state;

ρ_{dmin} is dry density in the loosest state;

ρ_{dmax} is dry density in the densest state.

式中，ρ_d 为正常状态下土的干密度，ρ_{dmin} 为最松散状态下土的干密度，ρ_{dmax} 为最密实状态下土的干密度。

According to Terzaghi, the conventional coefficients are as follows: $D_r = 0$, loosest; $D_r = 1$, densest; $D_r \leqslant 1/3$, loose; $1/3 < D_r \leqslant 2/3$, medium-dense; $D_r > 2/3$, dense.

根据太沙基的理论，$D_r = 0$ 代表土处于最松散状态；$D_r = 1$ 代表土处于最密实状态；$D_r \leqslant 1/3$ 代表土处于松散状态；$1/3 < D_r \leqslant 2/3$ 代表土处于中等密实状态；$D_r > 2/3$ 代表土处于密实状态。

Example 2. 4

There is a sample which is got from a certain natural sand column. The water content measured by experiment is 15%, bulk density is $\rho=1.80\mathrm{g/cm^3}$, the minimum dry density is $1.38\mathrm{g/cm^3}$, the maximum dry density is $1.65\mathrm{g/cm^3}$. Please determine the sandy soil′s dense degree.

例题 2. 4

从某个天然砂柱中取得的一个样本。通过试验测得其含水量为 15%，密度为 $\rho=1.80\mathrm{g/cm^3}$，最小干密度为 $1.38\mathrm{g/cm^3}$，最大干密度为 $1.65\mathrm{g/cm^3}$。确定这个砂样的密实度。

Solution

It is known that $\rho=1.80\mathrm{g/cm^3}$, the natural dry density of the sandy soil can be calculated according to

解答

已知 $\rho=1.80\mathrm{g/cm^3}$，可据此计算出砂土的干密度

$$\rho_{\mathrm{d}}=\frac{\rho}{1+w}=\frac{1.80}{1+0.15}=1.57\ \mathrm{g/cm^3}$$

and the using $\rho_{\min}=1.38\mathrm{g/cm^3}$, $\rho_{\max}=1.65\mathrm{g/cm^3}$ to equation.

然后将 $\rho_{\min}=1.38\mathrm{g/cm^3}$，$\rho_{\max}=1.65\mathrm{g/cm^3}$ 代入式（2. 27）中，得出 $D_{\mathrm{r}}=0.74$。

$$D_{\mathrm{r}}=\frac{(\rho_{\mathrm{d}}-\rho_{\mathrm{dmin}})\rho_{\max}}{(\rho_{\mathrm{dmax}}-\rho_{\mathrm{dmin}})\rho_{\mathrm{d}}}=\frac{(1.57-1.38)\times1.65}{(1.65-1.38)\times1.57}=0.74$$

Therefore, the sandy soil is in a dense state (because D_{r} is greater than 2/3).

所以砂土处于密实状态（因为 $D_{\mathrm{r}}=0.74$ 大于 2/3）。

2. 4. 2 Consistency of Cohesive Soil

Generally speaking, consistency is a characteristic of a material and is manifested through its resistance to flow. The term “Soil Consistency” conveys the idea of the degree of cohesion between the soil particles.

Also, consistency can be regarded as the outward result of the forces of cohesion acting at various degrees of water content. Figure 2. 4 shows the soil state changes from stiffer, soften to flow with the increase of water content.

2. 4. 2 黏性土的稠度

一般来说，稠度是材料的特性，它通过材料的抗流动性能表现出来。“土的稠度”表示土颗粒之间的黏聚程度。

此外，稠度也可以看作是在不同含水量下黏聚力作用的外在结果。图 2. 4 表示了随着含水量的增大，土的状态从较为坚硬到变软再到流动。

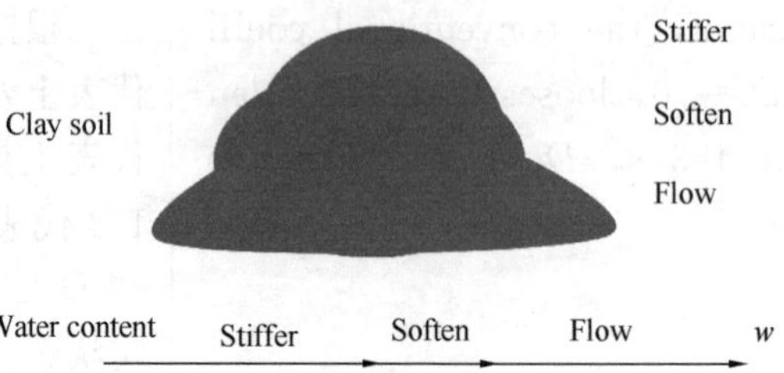

Figure 2. 4　The relationship between consistency state and water content

Consistency is commonly described by such terms as hard, plastic, soft or liquid. According to water content, cohesive soils are in different states. At a very low moisture content, soil behaves more like a solid. When the moisture content is very high, the soil may flow like a liquid. Hence, on an arbitrary basis, depending on the water content, the behavior of soil can be divided into four basic states-solid, semisolid, plastic, and liquid. The moisture content corresponding to the transition of cohesive soil from one physical state to another is called the boundary moisture content.

稠度通常用硬、塑性、软或液体等术语来描述。根据含水量的不同，黏性土处于不同的状态。在含水量很低的情况下，土会更像固体；当含水量非常高时，土可能会像液体一样流动。因此，根据含水量的不同，土可以分为四种基本状态：固态、半固态、塑性和液态。黏性土由一种物理状态转到另一种物理状态对应的含水量称为界限含水量。

Corresponding to the consistency state, water forms in soil shows in state of strong absorbed water, feeble absorbed water and free water. As shown in this Figure. Let′s explain the consistency limit in different water content. The soil transforms from the liquid into the plastic state named liquid limit with a symbol w_L.

土中水的形态与稠度状态相对应，表现为强结合水、弱结合水和自由水状态。下面解释不同含水量下的稠度极限。液限是土从液态变为塑性状态的界限，符号为 w_L。

Losing its ability to flow as a liquid state, the soil can be readily molded and hold its shape or can change shape without the appearance of cracks in it.

失去了液态流动的能力，土可以很容易地被塑造并保持其形状，或者可以在没有出现裂缝的情况下改变形状。

w_P is the lower limit of the plastic stage of soil. Upon a further decrease in water content, the plastic properties of the cohesive soil are lost, and at a certain water content termed the plastic limit up. The clay transforms from the plastic state into the semi-solid state. The water content of the soil at which the soil transforms from the semi-solid state to the solid state is termed the shrinkage limit w_s.

塑限 w_P 是土塑性阶段的下限。当土含水量进一步降低时，具有黏性的土的塑性特性会丧失，对应的含水量称为塑性限度。此时，黏土从塑性状态转变为半固态状态。缩限 w_s 是土从半固态状态转变为固态状态的含水量。

In China, the liquid limit is measured using a cone type liquid limit apparatus (Figure 2.5), and the plastic limit is measured using a liquid-plastic limit combined apparatus (Figure 2.6) . The test procedure can be refered in " Standard for Geotechnical Testing Methods" GB/T 50123—2019.

我国用锥式液限仪测定液限（图 2.5），可用液-塑限联合测定仪测定塑限（图 2.6）。其试验方法可参考《土工试验方法标准》GB/T 50123—2019。

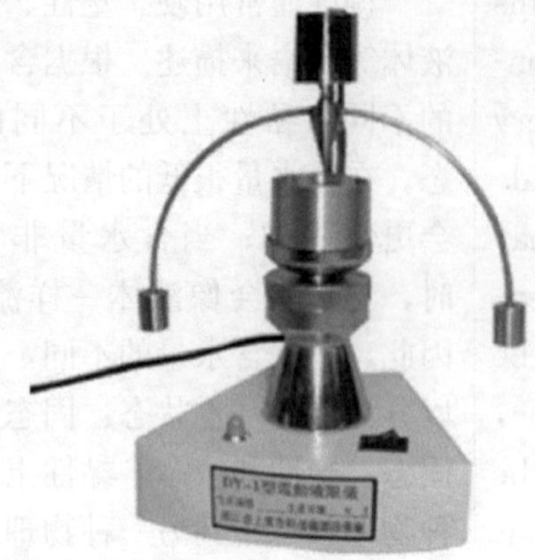
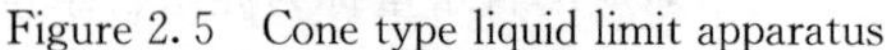

Figure 2.5 Cone type liquid limit apparatus

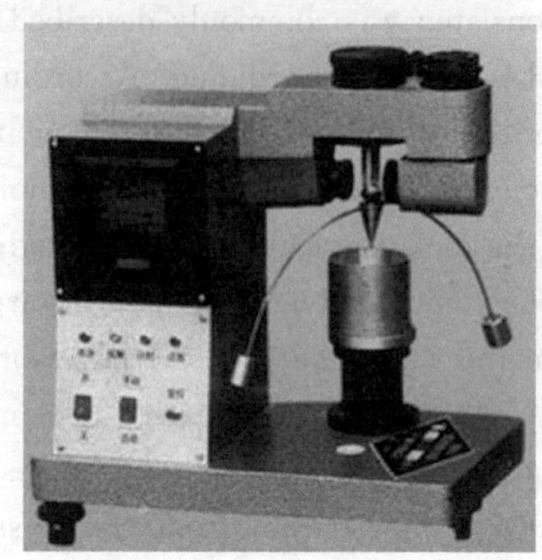

Figure 2.6 Liquid-plastic limit combined apparatus

In addition to the boundary moisture content, the physical state indicators of cohesive soil also include plasticity index and liquidity index.

黏性土的物理状态指标除界限含水量外，还有塑性指数和液性指数。

The difference in water content or interval between liquid and plastic limits is termed the plasticity index and can be expressed as:

含水量或液限和塑限之间的差异称为塑性指数，可表示为式(2.28)。

$$I_P = w_L - w_P \tag{2.28}$$

The plasticity index I_P represents the range of changes in the moisture content of soil in a plastic state.

塑性指数 I_P 表示土处在可塑状态的含水量变化范围。

Obviously, the larger the plasticity index, the wider the range of moisture content in which the soil is in a plastic state. The size of the plasticity index reflects the possible content of bound water in the soil. In terms of individual particle characteristics of soil, the finer the soil particles, the larger their specific surface area and the higher their bound water content, resulting in an increase in I_P. In terms of the mineral composition of soil particles, the higher the content of clay minerals (especially montmorillonite) particles, the more intense the hydration process, and the higher the bound water content, resulting in an increase in soil particles. In terms of the properties of aqueous solutions in soil, as the concentration of high valence cations in water increases, the number of cations adsorbed in the anti ion layer on the surface of soil particles decreases, and the thickness of the bound water layer

显然，塑性指数越大，土处于可塑状态的含水量范围也越大。塑性指数大小反映了土中结合水的可能含量。从土的颗粒特征个体而言，土粒越细，则其比表面积越大，结合水含量越高，I_P越大；从土的颗粒矿物成分而言，黏土矿物（尤以蒙脱石类）颗粒含量越高，水化作用剧烈，结合水含量越高，因而土的颗粒也越大；就土中水溶液性质而言，当水中高价阳离子的浓度增大时，土粒表面吸附的反离子层中阳离子数量减少，结合水层厚变薄，I_P随之减小；反之随着反离子层中的低价阳离子的增加，I_P增大。在一定程度上，塑性指

becomes thinner, resulting in a decrease in I_P. On the contrary, as the number of low-priced cations in the counter ion layer increases, I_P also increases. To a certain extent, the plasticity index I_P comprehensively reflects the basic characteristics of cohesive soil and its three-phase composition, and in engineering, cohesive soil is often classified according to the plasticity number I_P.

数 I_P综合反映了黏性土及其三相组成的基本特性，在工程上常按塑性指数 I_P 对黏性土进行分类。

The liquidity index refers to the ratio of the difference between the natural moisture content and plastic limit of cohesive soil to the plasticity index, represented by the symbol I_L. It can be expressed as:

液性指数是指黏性土的天然含水量和塑限的差值与塑性指数之比，用符号 I_L 表示。用公式可表示为式（2.29）。

$$I_L = \frac{w - w_P}{w_L - w_P} = \frac{w - w_P}{I_P} \tag{2.29}$$

The liquidity index I_L is mainly used to characterize the physical state of natural sedimentary cohesive soils. When the natural moisture content w of the soil is less than or equal to w_P, I_L is less than or equal to 0, and the natural soil is in a hard state; When w is greater than w_L, I_L is greater than 1, and the natural soil is in "flow state"; When w is between w_P and w_L, i. e. I_L is between 0 and 1, the natural soil is in a plastic state. Therefore, the liquidity index I_L can be used as a classification index for the state of cohesive soil. The higher the I_L value, the softer the soil; conversely, the harder the soil. Clay is classified into soft and hard states based on its liquidity index value, and the classification criteria are shown in Table 2. 2.

液性指数 I_L，主要用于表征天然沉积黏性土的物理状态。当土的天然含水率 w 小于等于 w_P时，I_L小于等于 0，天然土处于坚硬状态；当 w 大于 w_L时，I_L大于 1，天然土处于“流动态”；当 w 在 w_P与 w_L之间时，即 I_L为 0～1，则天然土处于可塑状态。因此，可以用液性指数 I_L作为黏性土状态的划分指标。I_L值越大，土质越软，反之，土质越硬。用液性指数值划分黏土软硬状态，划分标准见表 2. 2。

Physical state division of cohesive soil **Table 2. 2**

State	Hard	Hard plastic	Plastic	Soft plastic	Flow state
Liquidity index	$I_L \leqslant 0$	$0 < I_L \leqslant 0.25$	$0.25 < I_L \leqslant 0.75$	$0.75 < I_L \leqslant 1.0$	$I_L > 1.0$

It should be pointed out that in addition to sandy soil and cohesive soil, there is also a special type of soil in nature-silt. It is a transitional soil type between sandy soil and cohesive soil. The silt particles are mainly composed of extremely fine sand particles (0. 75～0. 1mm) and silt particles (0. 005～0. 075mm), which have unique physical and mechanical properties, possessing both sandy and cohesive soil properties.

需要指出，除砂类土和黏性土外，自然界中还存在一种特殊土类——粉土。它是介于砂土和黏性土之间的过渡土类。粉土颗粒以极细砂粒（0. 075～0. 1mm）和粉粒（0. 005～0. 075mm）粒组为主，其物理力学性质较为特殊，既具有砂土的性质又具有黏性土的性质。

2.5 Soil Description and Classification

It is essential that a standard language should exist for the description of soils. A comprehensive description should include the characteristics of both the soil material and the in-situ soil mass. Material characteristics can be determined from disturbed samples of the soil, i. e. samples having the same particle size distribution as the in-situ soil but in which the in-situ structure has not been preserved.

The principal material characteristics are particle size distribution (or grading) and plasticity, from which the soil name can be deduced. Particle size distribution and plasticity properties can be determined either by standard laboratory tests or by simple visual and manual procedures. Secondary material characteristics are the color of the soil and the shape, texture and composition of the particles. Mass characteristics should ideally be determined in the field, but in many cases, they can be detected in undisturbed samples, i. e. samples in which the in-situ soil structure has been essentially preserved.

A description of mass characteristics should include an assessment of in-situ compactive state (coarse-grained soils) or stiffness (fine-grained soils), and details of any bedding, discontinuities and weathering. The arrangement of minor geological details, referred to as the soil macro-fabric, should be carefully described, as this can influence the engineering behavior of the in-situ soil to a considerable extent. Examples of macro-fabric features are thin layers of fine sand and silt in clay, silt-filled fissures in clay, small lenses of clay in sand, organic inclusions, and root holes. The name of the geological formation, if definitely known, should be included in the description; in addition, the type of deposit may be stated (e. g. till, alluvium), as this can indicate, in a general way, the likely behavior of the soil.

It is important to distinguish between soil descrip-

2.5 土的描述和分类

必须有一种标准语言来描述土。综合描述应包括土的材料和原位土的特征。材料特征可以由土的扰动试样确定，即与原位土具有相同粒径分布但未保留原位结构的试样。

主要材料特征有颗粒分布（或级配）和塑性，从中可以推断出土的名称。粒径分布和塑性性能可以通过标准实验室试验或简单的目测和手动试验来确定。次要材料特征是指土的颜色以及颗粒的形状、质地和成分。理想情况下，质量特征应在现场确定，但在许多情况下，可用未受干扰的试样（即基本保留了原位土的结构的试样）检测。

对质量特征的描述应包括对现场密实状态（粗粒土）或刚度（细粒土）的评估，以及任何层理、不连续性和风化的细节。应仔细描述被称为土宏观组构的次要地质细节的排列，因为这会在相当大的程度上影响现场土的工程行为。宏观组构特征包括黏土中细砂和粉土的薄层、黏土中粉土填充的裂缝、砂中黏土的小透镜体、有机包裹体和根孔等。如果明确知道地质构造的名称，应补充在描述中；此外，可以说明沉积物的类型（如冲积层、冲积层），因为这可以概括地表明土的可能行为。

区分土的描述和土的分类很

tion and soil classification. Soil description includes details of both material and mass characteristics, and therefore it is unlikely that any two soils will have identical descriptions. In soil classification, a soil is allocated to one of a limited number of behavioral groups on the basis of material characteristics only. Soil classification is thus independent of the in-situ condition of the soil mass.

重要。土的描述包括材料和质量特性的详细信息，因此任何两种土都不太可能有相同的描述。在土的分类中，仅根据材料特征就能将土划分到某组。因此，土的分类与土体的原位条件无关。

2.5.1 General System of Soil Classification

There are numerous types of soil in nature with different engineering properties. The classification system of soil is to classify soil into certain categories based on its engineering properties. Its purpose is to use a universal identification standard to facilitate valuable comparisons, evaluations, and academic research and engineering applications among different soil types. Currently, the soil classification system used in engineering is shown in Figure 2.7.

2.5.1 土的分类系统

自然界中有许多类型的土，具有不同的工程性质。土的分类系统是根据土的工程性质将土分为某些类别。其目的是使用通用的识别标准，以实现不同类型的土之间的比较、评估、学术研究和工程应用。目前，工程中使用的土的分类系统如图 2.7 所示。

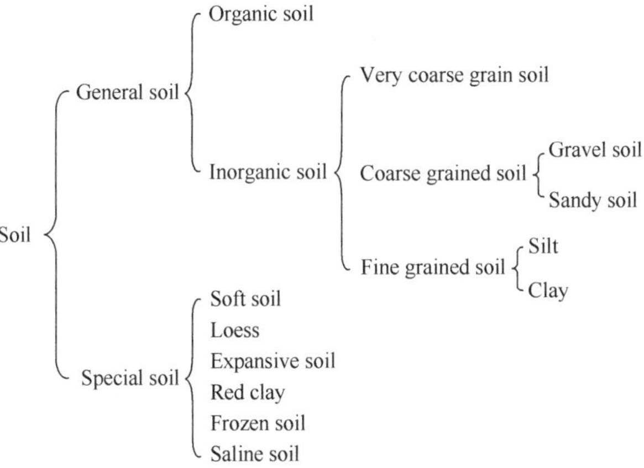

Figure 2.7 Soil classification system

In 2007, China revised the " Standard for Engineering Classification of Soil" GB/T 50145—2007. According to the range of soil particle size, soil is divided into three major particle groups: very coarse grain, coarse grain, and fine grain. It is further divided into six major particle groups: boulder or block stone grain, cobble or broken stone grain, gravel grain, sand, silt, and clay. The division of particle

2007 年，中国修订了《土的工程分类标准》GB/T 50145—2007。根据土的粒径的范围，土被分为三大颗粒组：巨粒、粗粒和细粒。它进一步分为六大粒组：漂石或块石、卵石或碎石，砾粒、砂粒、粉粒和黏

group is shown in Table 2. 3. The classification criteria for coarse-grained soil and fine-grained soil are shown in Table 2. 4 and Table 2. 5, respectively.

粒。颗粒划分见表 2. 3。粗粒土和细粒土的分类标准分别见表 2. 4 和表 2. 5。

Division of particle group **Table 2. 3**

Generic terms of soil particles	Name of grain group	Range of particle diameter d (mm)
Very coarse grain	Boulder (block stone) grain	$d > 200$
	Cobble (broken stone) grain	$200 \geqslant d > 60$
Coarse grain	Gravel grain	$60 \geqslant d > 2$
	Sand	$2 \geqslant d > 0.075$
Fine grain	Silt	$0.075 \geqslant d > 0.005$
	Clay	$d \leqslant 0.005$

Division of very coarse particle soil and soil containing very coarse particle **Table 2. 4**

Soil type	Content of grain group		Group symbols	Soil's name
Very coarse grain soil	75% < Content of very coarse grain < 100%	Content of boulders >content of cobbles	B	Boulder
		Content of boulders ⩽content of cobbles	Cb	Cobble
Mixed very coarse grain soil	50% < Content of very coarse grain ⩽ 75%	Content of boulders >content of cobbles	BSI CbSI	Boulders soil
		Content of boulders ⩽content of cobbles	BSI CbSI	Cobbles soil
Very coarse grain mixed soil	15% < Content of very coarse grain ⩽ 50%	Content of boulders > content of cobbles	SIB	Soil with boulders
		Content of boulders ⩽ content of cobbles	SICb	Soil with cobbles

Division of fine-grained soil (cone depth 17mm in liquid limit test) **Table 2. 5**

Plastic index I_P	Liquid limit w_L (%)	Soil's code	Soil's name
$I_P \geqslant 0.73\ (w_L - 20)$ and $I_P \geqslant 10$	$w_L \geqslant 50$	CH	High liquid limit clay
	$w_L < 50$	CL	Low liquid limit clayl
$I_P < 0.75\ (w_L - 20)$ or $I_P < 10$	$w_L \geqslant 50$	MH	High liquid limit silt
	$w_L < 50$	ML	Low liquid limit silt

2.5.2 Soil Classification in Building Foundation

The main feature of the classification system of the " Code for Design of Building Foundation" GB 50007—2011 and the " Code for Investigation of Geotechnical Engineering" GB 50021—2001 (2009 edition) is that it focuses on the natural structural characteristics and strength of soil, and is always closely related to the main engineering characteristics of soil—deformation and strength characteristics. Therefore, the division based on sedimentary age and geological genesis was first considered, while distinguishing regional special soils with certain special formation conditions and engineering properties from ordinary soils.

2.5.2 建筑地基土的分类

《建筑地基基础设计规范》GB 50007—2011 和《岩土工程勘察规范》GB 50021—2001（2009 年版）的分类体系的主要特征是：它侧重于土的自然结构特性和强度，并始终与土的主要工程特性——变形和强度特性密切相关，因此，首先考虑了基于沉积时代和地质成因的划分，同时将具有特定形成条件和工程性质的区域特殊土与普通土区分开来。

1. Divided by sedimentary age and geological genesis

The foundation soil can be divided into: ① Old sedimentary soil: soil deposited before Q_3 of the Late Pleistocene in the Quaternary period, generally in an over consolidated state with high structural strength; ② Newly deposited soil: Soil deposited recently in the Quaternary Holocene, generally in an under consolidated state with low structural strength.

According to geological genesis, soil can be divided into residual soil, slope soil, alluvial soil, lacustrine soil, marine soil, aeolian soil, and glacial soil.

1. 按沉积年代和地质成因划分

地基土可分为：①老沉积土：第四纪晚更新世 Q_3 及以前沉积的土，一般处于超固结状态，结构强度高；②新近沉积土：第四纪全新世近期沉积的土，通常处于欠固结状态，结构强度低。

根据地质成因土可分为残积土、坡积土、冲积土、湖积土、海积土、风积土和冰积土。

2. Divided by particle size distribution (particle size composition) and plasticity index

Soil is divided into four categories based on particle size distribution and plasticity index: gravel soil, sandy soil, silty soil, and cohesive soil.

(1) Gravel soil

Soil with a particle size greater than 2mm and a content exceeding 50% of its total weight is called gravel soil. According to the particle size distribution and shape, it is divided into boulders, block stones, pebbles, crushed stones, round gravel, and angular gravel are shown in Table 2.6.

2. 按颗粒级配（粒度成分）和塑性指数划分

根据颗粒级配和塑性指数，土可分为碎石土、砂土、粉土和黏性土四大类。

(1) 碎石土

粒径大于 2mm 颗粒含量超过其总重量 50%的土称为碎石土。根据颗粒级配和颗粒形状按表 2.6 分为漂石、块石、卵石、碎石、圆砾和角砾。

Division of gravel soil（GB 50007—2011） **Table 2.6**

Soil's name	Shape of particle	Particle size distribution
Boulder grain	Mainly Circular and sub-circular shape	Particle content ($d>$ 200mm) $>$ 50% Total weight
Block stone	Angular shape	
Pebbles	Mainly Circular and sub-circular shape	Particle content ($d>$ 20mm) $>$ 50% Total weight
rushed stones	Angular shape	
Round gravel	Mainly Circular and sub-circular shape	Particle content ($d>$ 2mm) $>$ 50% Total weight
Angular gravel	Mainly Angular shape	

(2) Sandy soil

Soil with a particle size greater than 2mm and a particle size greater than 0.075mm，which accounts for no more than 50% and more than 50% of the total weight，respectively，is called sandy soil. According to the particle size distribution，it is divided into gravel sand，coarse sand，medium sand，fine sand，and silt are shown in Table 2.7.

（2）砂土

粒径大于 2mm 的颗粒含量不超过全重的 50%，且粒径大于 0.075mm 的颗粒含量超过全重的 50%的土称为砂土。根据颗粒级配按表 2.7 分为砾砂、粗砂、中砂、细砂和粉砂。

Division of sandy soil（GB 50007—2011） **Table 2.7**

Soil's name	Particle size distribution
Gravel sand	25% $<$ Particle content ($d>$ 2mm) $<$ 50%
Coarse sand	Particle content ($d>$ 0.5mm) $>$ 50%
Medium sand	Particle content ($d>$ 0.25mm) $>$ 50%
Fine sand	Particle content ($d>$ 0.075mm) $>$ 85%
Silt	Particle content ($d>$ 0.075mm) $>$ 50%

(3) Silty soil

Silty soil is between sandy soil and cohesive soil，with a plasticity index of $I_P \leqslant 10$，and a particle size greater than 0.075mm，with a content of no more than 50% of the total weight of the soil. Generally，according to regional regulations (such as Shanghai，Tianjin，Shenzhen，etc.)，based on the content of clay particles，it can be further divided into clayey silt and sandy silt according to Table 2.8.

（3）粉土

粉土介于砂土和黏性土之间，塑性指数 $I_P \leqslant 10$，粒径大于 0.075mm 的颗粒含量不超过土全重的 50%。一般按地区规定（如上海、天津、深圳等），按黏粒含量可按表 2.8 进一步分为黏质粉土和砂质粉土。

Division of silty soil **Table 2.8**

Soil's name	Particle size distribution
Sandy silt	Particle content($d < 0.005$mm) $\leqslant 10\%$
Clayey silt	Particle content ($d < 0.005$mm) $> 10\%$

(4) Cohesive soil

Soil with a plasticity index greater than 10 is called cohesive soil. According to the plasticity index I_P, it can be further divided into silty clay and clay according to Table 2.9.

(4) 黏性土

塑性指数大于10的土称为黏性土。根据塑性指数 I_P 按表2.9进一步分为粉质黏土和黏土。

Division of cohesive soil (GB 50007—2011) **Table 2.9**

Soil's name	Plasticity index I_P
Silty clay	$10 < I_P \leqslant 17$
Clay	$I_P > 17$

Note: The plasticity index is calculated from the liquid limit measured when the corresponding 76g cone sinks into the soil sample at a depth of 10mm.

注：塑性指数由相应76g圆锥体沉入土样中深度为10mm时测定的液限计算而得。

3. Special soil

Soil with a certain distribution area or engineering significance and special composition, state, and structural characteristics is called special soil. It is divided into collapsible soil, adamic earth (red soil), soft soil (including silt, silty soil, peaty soil, peat, etc.), mixed soil, filled soil, perennially frozen soil, expansive rock soil, saline rock soil, weathered rock and residual soil, and polluted soil. See "Code for Investigation of Geotechnical Engineering" GB 50021—2001 (2009 edition) for details.

It is worth noting that with the deepening of China's lunar exploration project, people are paying more attention to the study of lunar soil. It refers to the loose particle accumulation material covering the bedrock on the surface of the moon.

The formation process of lunar soil is different from that of earth soil. Lunar soil is ultimately formed under the mechanical effects of frequent impacts of meteorites and micrometeorites, accumulation of impact ejecta, bombardment by cosmic rays and solar wind, and intense temperature differences that cause thermal expansion, contraction, and fragmentation of lunar surface rocks.

3. 特殊土

具有一定分布区域或工程意义和特殊成分、状态和结构特征的土称为特殊土。分为湿陷性土、红黏土、软土（包括淤泥、淤泥质土、泥炭土、泥炭等）、混合土、填土、多年冻土、膨胀岩土、盐渍岩土、风化岩和残积土、污染土。详见《岩土工程勘察规范》GB 50021—2001（2009年版）。

值得注意的是，随着我国探月工程的深入，人们对月壤的研究也愈加关注。月壤是指覆盖在月球表面基岩以上的松散颗粒堆积物质。

月球土壤与地球土壤的形成过程不同，月壤是在陨石和微陨石的频繁撞击、撞击溅射物堆积、宇宙射线和太阳风的轰击，以及剧烈的温差促使月表岩石热胀冷缩破碎等机械作用下最终形成。

The lunar samples brought back by the United States' Apollo and China's Chang'e 5 and 6 missions indicate that lunar soil is primarily a mixture of three types of materials: highland rocks and breccias, KREEP (potassium, rare earth elements, and phosphorus) rocks, and mare basalts and volcanic glass. The chemical composition, rock types, and mineral composition of lunar soil are complex, with each lunar soil sample consisting of various rocks and minerals. The particle size distribution of lunar soil is very wide, with particle diameters mainly less than 1.0mm, and the vast majority of particle diameters ranging from 30μm to 1.0mm. The particle morphology of lunar soil is extremely variable, ranging from spherical and ellipsoidal to extremely angular shapes. Elongated, sub-angular, and angular particle shapes are relatively more common. Generally, clumps with a diameter of ≥10mm are treated and studied as rocks, referred to as Lunar Rocks; particles with a diameter of <10mm are considered Lunar Soil in the narrow sense; and Lunar Dust (or Lunar Fines) refers to particles with a diameter of <1.0mm within the Lunar Soil. The natural density of lunar soil generally increases with depth, with the density of the lunar surface soil being approximately 1.15 to 1.55g/cm^3. Human understanding and research of lunar soil are essential foundations for lunar exploration, as well as for future endeavors such as establishing lunar bases and lunar colonization. The mechanical indices of lunar soil, such as its compressibility and shear strength, urgently require in-depth study.

美国阿波罗及我国嫦娥五号、六号带回的月球样品表明，月壤主要是高地岩石和角砾岩、克里普岩（KREEP）以及月海玄武岩和火山玻璃三类物质的混合物。月壤的化学成分、岩石类型和矿物组成复杂，每一个月壤样品都由多种岩石和矿物组成。月壤粒度分布很广，颗粒直径以小于1.0mm为主，绝大部分颗粒直径在30μm～1.0mm之间。月壤的颗粒形态极为多变，从圆球状、椭球状到极端棱角状都有。长条状、次棱角状和棱角状的颗粒形态相对更为常见。一般把直径≥10mm的团块当作岩石对待，进行处理和研究，称为月岩（Lunar Rocks）；直径<10mm的颗粒才是狭义上的月壤（Lunar Soil）；而月尘（Lunar Dust或Lunar Fines）则是指月壤中直径<1.0mm的颗粒。月壤天然密度一般随深度的增加而增大，月球表层月壤密度约1.15～1.55g/cm^3。人类对月壤的认识和研究是月球探测，以及未来建立月球基地、移民月球等不可或缺的基础，月壤的压缩性、抗剪强度等力学指标亟待深入研究。

Additionally, deep-sea energy soil, as a type of seabed sediment containing natural gas hydrates, differs from terrestrial soil. It is a new type of clean energy and also classified as a special soil. It features wide distribution, deep burial, pollution-free, and high energy density, with extremely broad development prospects. The understanding of the physical and mechanical properties and engineering characteristics of energy soil is still immature. The author's research team has conducted related studies on the static and dynamic properties of deep-sea energy soil, see references [15-17].

另外，深海能源土作为一种含天然气水合物的海底沉积物，不同于陆地土，是一种新型清洁能源，亦属于特殊土。它具有分布广、埋藏深、无污染、能量密度高等特点，其开发前景极为广阔。人们对能源土物理力学及工程性质的认识尚未成熟，编者课题组对深海能源土的静、动力学性质开展了相关研究，参见文献[15-17]。

Problems

2.1 For a soil with $d_{60}=0.42$mm, $d_{30}=0.21$mm, and $d_{10}=0.16$mm, calculate uniformity coefficient and the coefficient of gradation.

2.2 The following are the result of a sieve analysis:

U. S. sieve no.	Mass of soil retained (g)
4	28
10	42
20	48
40	128
60	221
100	86
200	40
Pan	24

a. Determine the percent finer than each sieve and plot a grain-size distribution curve.

b. Determine d_{10}, d_{30}, and d_{60} from the grain-size distribution curve.

c. Calculate the non-uniformity coefficient, C_u.

d. Calculate the curvature coefficient, C_c.

2.3 Undisturbed soil sample was collected from the field in steel Shelby tubes for laboratory evaluation. The tube sample has a diameter of 71 mm, length of 558 mm, and a moist weight of 42.5×10^{-3} kN. If the oven-dried weight was 37.85×10^{-3} kN, and $G_s=2.69$, calculate the following:

a. Moist unit weight.

b. Field moisture content.

c. Dry unit weight.

d. Void ratio.

e. Degree of saturation.

2.4 For a given soil, the following are known: $G_s=2.74$, moist unit weight, $\gamma=20.6\text{kN/m}^3$, and moisture content, $w=16.6\%$. Determine:

a. Dry unit weight.

b. Void ratio.

c. Porosity.

d. Degree of saturation.

2.5 For a moist soil, given the following: $V=7.08\times10^{-3}\ \text{m}^3$; $W=136.8\times10^{-3}$ kN; $w=9.8\%$; $G_s=2.66$. Determine:

a. Dry unit weight.

b. Void ratio.

c. Volume occupied by water.

2.6 The moisture content of a soil sample is 17% and the dry unit weight is 16.51 kN/m^3. If $G_s = 2.69$, what is the degree of saturation?

2.7 For a given sandy soil, $e_{max} = 0.75$ and $e_{min} = 0.52$. If $G_s = 2.67$ and $D_r = 65\%$, determine:

a. Void ratio.

b. Dry unit weight.

2.8 For a given sandy soil the maximum and minimum void ratios are 0.72 and 0.46, respectively. If $G_s = 2.68$ and $w = 11\%$, what is the moist unit weight of compaction (kN/m^3) in the field if $D_r = 82\%$?

Chapter 3 Permeability and Seepage

第 3 章 渗透性与渗流

3.1 Introduction

Any given mass of soil consists of solid particles of various sizes with inter connected void spaces. The continuous void spaces in a soil permit water to flow from a point of high energy to a point of low energy. The flow of groundwater in soil pores is called seepage, and the characteristic of soil that can be penetrated by water is defined as permeability.

The issues related to soil permeability in engineering mainly include three aspects:

(1) Seepage flow problem: calculation of water inflow in foundation pits or construction cofferdams; The seepage volume of the canal dam body and foundation. (Figure 3.1)

(2) Penetration failure problem: The collapse of Teton Dam and Dongting Lake Dam are both caused by the movement of soil particles or soil due to the seepage force generated by water flowing in soil pores, leading to infiltration failure. (Figure 3.2)

(3) Seepage control problem: In engineering, when the seepage flow rate or seepage deformation does not meet the design requirements, engineering measures must be taken to control seepage. (Figure 3.3)

3.1 概述

土都是由各种尺寸的固体颗粒组成的，这些颗粒之间有相互连通的孔隙。土中连续的孔隙允许水从高能点流向低能点。地下水在土孔隙中的流动称为渗流，土体的这种能够被水透过的特性称为渗透性。

工程中与土的渗透性有关的问题主要包括 3 个方面：

（1）渗流量问题：基坑或施工围堰的涌水量计算；水渠坝身、坝基渗水量（图 3.1）。

（2）渗透破坏问题：美国 Teton 坝溃坝和洞庭湖溃坝都是由于水在土孔隙中流动产生的渗流力引起土颗粒或土体的移动而诱发渗透破坏造成的（图 3.2）。

（3）渗流控制问题：工程中当渗流量或渗透变形不满足设计要求时，就要采取工程措施进行渗流控制（图 3.3）。

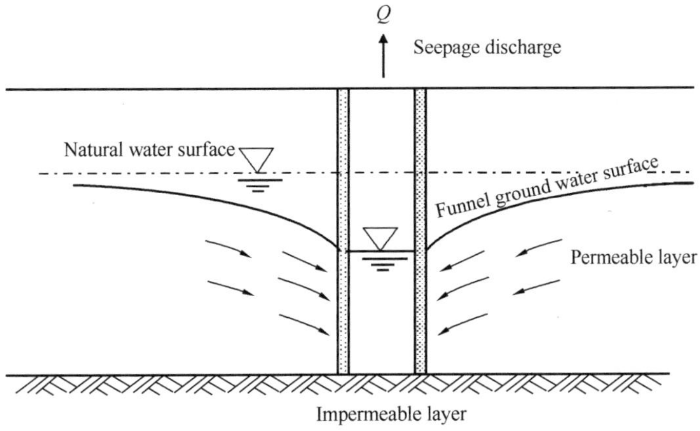

Figure 3.1 Well seepage

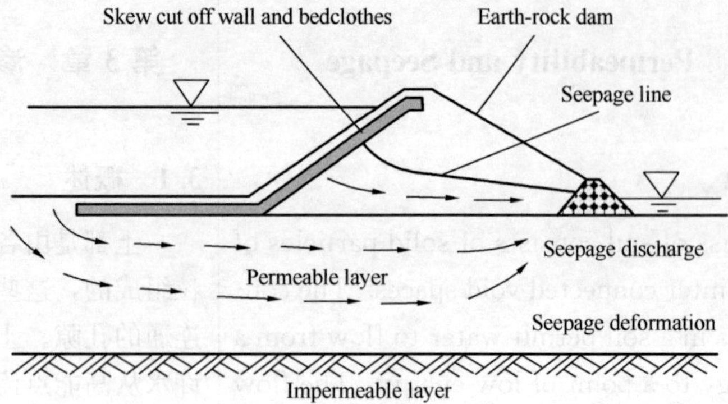

Figure 3.2　Seepage in the dam body and foundation

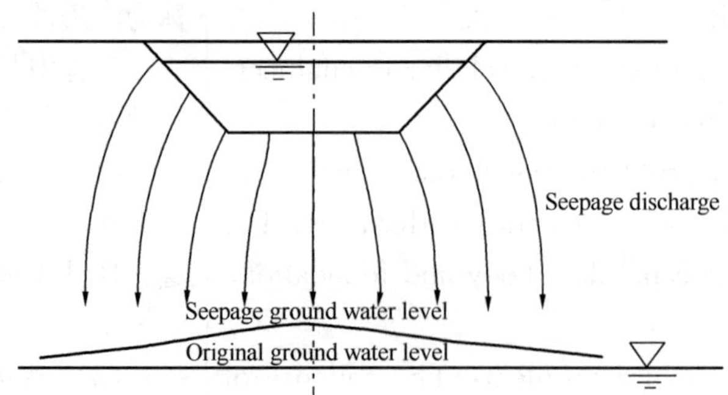

Figure 3.3　Canal seepage

One of the major physical parameters of soil that controls through the coefficient of permeability. In this chapter, we will study the following:

(1) Definition of hydraulic conductivity and its magnitude in various soils.

(2) Laboratory determination of coefficient of permeability.

(3) Equivalent coefficient of permeability in stratified soil based on the direction of the flow of water.

(4) coefficient of permeability determination from field tests.

土的主要物理参数之一是渗透系数。在本章中，我们将学习以下内容：

（1）不同土中渗透系数的定义及其大小。

（2）实验室测定渗透系数。

（3）基于水流方向的分层土等效渗透系数。

（4）通过现场试验测定渗透系数。

3.2　Fundamentals of Seepage

From fluid mechanics, we know that, according to Bernoulli's equation, the total head at a point in water under motion can be given by the sum of the pressure, velocity, elevation heads, or be expressed as:

3.2　渗流基础

根据流体力学中的伯努利方程，流动的水中一点的总水头可表达为压力水头、速度水头和位置水头的和，可表示为式(3.1)。

$$h = \frac{u}{\gamma_w} + \frac{v^2}{2g} + z \tag{3.1}$$

where h = total head;

u = pressure;

v = velocity;

g = acceleration due to gravity;

γ_w = unit weight of water.

式中，h 为总水头；u 为压力；v 为流速；g 为重力加速度；γ_w 为水的重度。

Note that the elevation head, z, is the vertical distance of a given point above or below a datum plane. The pressure head is the water pressure, u, at that point divided by the unit weight of water, γ_w.

设位置水头 z 是基准面上方或下方给定点的垂直距离。压力水头是该点的水压力 u 除以水的重度 γ_w。

If Bernoulli's equation is applied to the flow of water through a porous soil medium, the term containing the velocity head can be neglected because the seepage velocity is small, and the total head at any point can be adequately represented by:

如果将伯努利方程应用于通过多孔介质的水流，则包含速度水头的项可以忽略不计，因为渗流速度很小，任何点的总水头都可以用式（3.2）表示。

$$h = \frac{u}{\gamma_w} + z \tag{3.2}$$

Figure 3.4 shows the relationship among pressure, elevation, and total heads for the flow of water through soil. Open standpipes called piezometers are installed at points A and B. The levels to which water rises in the piezometer tubes situated at points A and B are known as the piezometric levels of points A and B, respectively. The pressure head at a point is the height of the vertical column of water in the piezometer installed at that point.

图 3.4 显示了土中水的压力、高程和总水头之间的关系。在 A 点和 B 点安装了开放式立管测压计。位于 A 点和 B 点的测压管中的水位分别称为 A 点和 B 点的测压计水位。某点的压力水头即安装在该点的压力计中垂直水柱的高度。

The loss of head between two points, A and B, can be given by:

A 和 B 两点之间的水头损失可由式（3.3）算出。

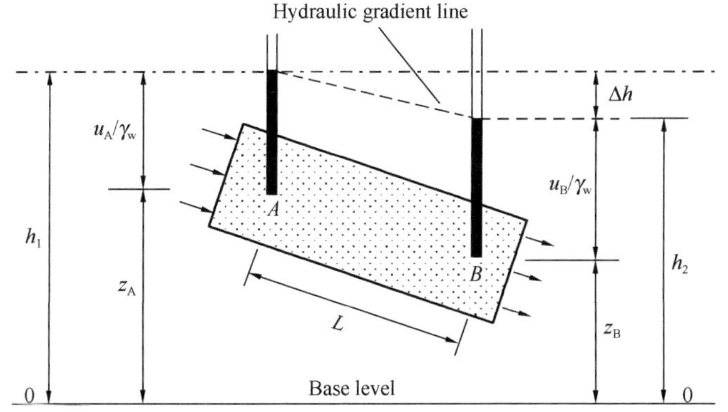

Figure 3.4 Pressure, elevation, and total heads for the flow of water

$$\Delta h = h_A - h_B = \left(\frac{u_A}{\gamma_w} + z_A\right) - \left(\frac{u_B}{\gamma_w} + z_B\right) \tag{3.3}$$

The head loss, Δh, can be expressed in a nondimensional form as:

水头损失 Δh 可以无量纲形式表示为式（3.4）。

$$i = \frac{\Delta h}{L} \tag{3.4}$$

3.3 Darcy's Law

3.3 达西定律

In 1856, Darcy published a simple equation for the discharge velocity of water through saturated soils, which may be expressed as:

1856 年，达西发表了一个关于水通过饱和土的渗流速度的简单方程，可以表示为式（3.5）。

$$v = ki \tag{3.5}$$

where v is discharge velocity, which is the quantity of water flowing in unit time through a unit gross cross-sectional area of soil at right angles to the direction of flow; k is hydraulic conductivity (otherwise known as the coefficient of permeability).

式中，v 为水的流速，即单位时间内流过与流向垂直的土单位总横截面积的水量；k 为渗透系数。

This equation was based primarily on Darcy's observations about the flow of water through clean sands. Note that it is valid for laminar flow conditions and applicable for a wide range of soils.

这个方程式主要基于达西对水在干净砂土中流动的观察。请注意，它适用于层流条件下的大部分土类。

In Eq. (3.5), v is the discharge velocity of water based on the gross cross-sectional area of the soil. However, the actual velocity of water (that is, the seepage velocity) through the void spaces is greater than v. A relationship between the discharge velocity and the seepage velocity can be derived by referring to Figure 3.5, which shows a soil of length L with a gross cross-sectional area A. If the quantity of water flowing through the soil in unit time is q, then

式（3.5）中，v 是基于土总横截面积的水的排放速度。然而，通过孔隙的实际水流速度（即渗流速度）大于 v。参考图 3.5 可以得出排水速度和渗流速度之间的关系，图 3.5 显示了长度为 L、总横截面面积为 A 的土样。如果单位时间内流过土的水量为 q，则可得式（3.6）：

$$q = vA = A_v v_s \tag{3.6}$$

where v_s is seepage velocity, A_v is area of void in the cross section of the sample.

式中，v_s是渗流速度，A_v是试样横截面中孔隙的面积。

However,

然而，土总横截面面积等于土固体面积和孔隙面积之和［式（3.7）］，

$$A = A_v + A_s \tag{3.7}$$

where A_s is area of soil solids in the cross section of the

式中，A_s是试样横截面中固体

sample.

Combining Eqs. (3.6) and (3.7) gives

颗粒面积。

由式(3.6)和式(3.7)可得式(3.8)或式(3.9),

$$q = v(A_v + A_s) = A_v v_s \tag{3.8}$$

or

$$v_s = \frac{v(A_v + A_s)}{A_v} = \frac{v(A_v + A_s)L}{A_v L} = \frac{v(V_v + V_s)}{V_v} \tag{3.9}$$

where V_s and V_v are volume of soil solids and voids in the sample respectively.

式中,V_s和V_v分别为试样的固体体积和孔隙体积。而式(3.9)可进一步写为式(3.10),

$$v_s = v\left[\frac{1+\left(\frac{V_v}{V_s}\right)}{\frac{V_v}{V_s}}\right] = v\left(\frac{1+e}{e}\right) = \frac{v}{n} \tag{3.10}$$

where e is void ratio, n is porosity.

式中,e和n分别为土样的孔隙比和孔隙率。

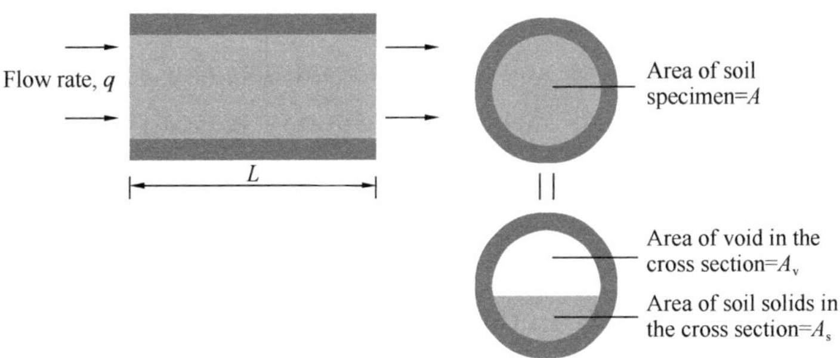

Figure 3.5 Derivation of Eq. (3.9)

Darcy's law as defined by Eq. (3.5) implies that the discharge velocity v bears a linear relationship to the hydraulic gradient i and passes through the origin as shown in Figure 3.6. However, Hansbo reported the test results for four undisturbed natural clays. On the basis of his results, a hydraulic gradient i' (Figure 3.6) appears to exist, at which

式(3.5)定义的达西定律意味着渗流速度v与水力梯度i呈线性关系,并过原点,如图3.6所示。然而,汉斯博报告了四种未受干扰的天然黏土的试验结果。根据他的结果,存在水力梯度i'(图3.6),可表达为式(3.11)和式(3.12)。

$$v = k(i - i_0) \qquad (\text{for } i \geqslant i') \tag{3.11}$$

.and

$$v = ki^m \qquad (\text{for } i < i') \tag{3.12}$$

The preceding equation implies that for very low hydraulic gradients, the relationship between v and i is nonlinear. The value of m in Eq. (3.12) for clay was about 1.5.

上述公式表明,对于非常低的水力梯度,v和i之间的关系是非线性的。对于黏性土,式(3.12)中的m值约为1.5。

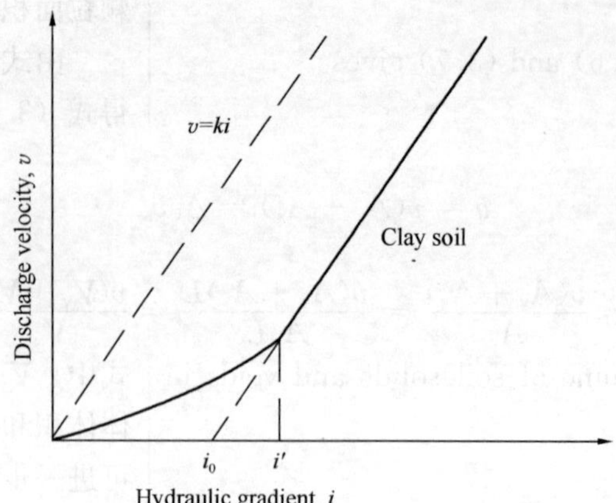

Figure 3.6 Variation of discharge velocity with hydraulic gradient in clay

3.4 Laboratory Determination of Coefficient of Permeability

Two standard laboratory tests are used to determine the hydraulic conductivity of soil—the constant-head test and the falling-head test. A brief description of each follows.

1. Constant-Head Test

A typical arrangement of the constant-head permeability test is shown in Figure 3.7. In this type of laboratory setup, the water supply at the inlet is adjusted in such a way that the difference of head between the inlet and the outlet remains constant during the test period. After a constant flow rate is established, water is collected in a graduated flask for a known duration.

The total volume of water collected may be expressed as:

$$Q = Avt = A(ki)t \tag{3.13}$$

where Q is volume of water collected;

A is area of cross section of the soil sample;

t is duration of water collection.

And because

$$i = \frac{h}{L} \tag{3.14}$$

where L = length of the sample, Eq. (3.14) can be substituted into Eq. (3.13) to yield

3.4 渗透系数室内试验测定

常通过常水头和变水头两种标准试验测定渗透系数。下面简要描述这两种方法。

1. 常水头试验

典型的常水头渗透试验装置如图 3.7 所示。在这种装置中，通过调整入口供水实现试验期间入口和出口的水头差恒定。设置恒定流速后，在持续时间内将水收集在刻度瓶中。收集的总水量表示为式（3.13）。

式中，Q 为收集的总水量，A 为土的横截面面积，t 为集水时间。

式中，L 为试样长度。将式（3.14）代入（3.13）可得式（3.15）或式（3.16）。

$$Q = A\left(k\frac{h}{L}\right)t \tag{3.15}$$

or

$$k = \frac{QL}{Aht} \tag{3.16}$$

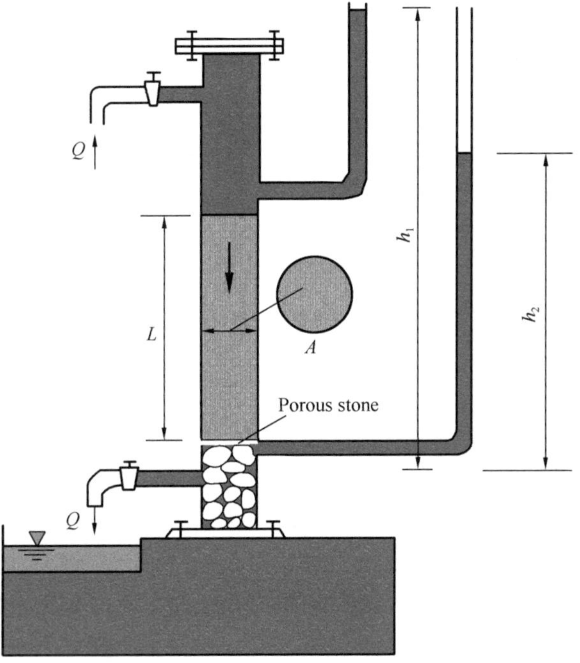

Figure 3.7 Constant-head permeability test

2. Falling-Head Test

A typical arrangement of the falling-head permeability test is shown in Figure 3.8. Water from a standpipe flows through the soil. The initial head difference h_1 at time $t = 0$ is recorded, and water is allowed to flow through the soil sample such that the final head difference at time $t = t_2$ is h_2.

The rate of flow of the water through the sample at any time t can be given by:

2. 变水头试验

典型变水头渗透试验装置如图 3.8 所示。水从立管内流进土样，记录 $t=0$ 时的初始水头差 h_1，水流过土样后，记录 $t=t_2$ 时的最终水头差 h_2。

在任意时刻 t，水通过土样的流速表示为式（3.17）。

$$q = k\frac{h}{L}A = -a\frac{dh}{dt} \tag{3.17}$$

Where q = flow rate;

a = cross-sectional area of the standpipe;

A = cross-sectional area of the soil sample.

Rearrangement of Eq. (3.17) gives

式中，q 为流速，a 为立管横截面面积，A 为土样的横截面面积。

由式（3.17）可得式（3.18）。

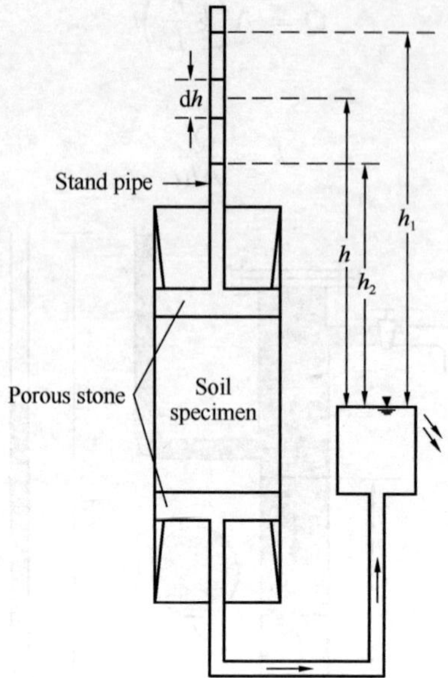

Figure 3.8　Falling-head permeability test

$$\mathrm{d}t = \frac{aL}{Ak}\left(-\frac{\mathrm{d}h}{h}\right) \tag{3.18}$$

Integration of the left side of Eq.（3.18）with limits of time from 0 to t and the right side with limits of head difference from h_1 to h_2 gives

将式（3.18）的左侧按时间从 0 到 t 积分，右侧按水头差从 h_1 到 h_2 积分，得式（3.19）。

$$t = \frac{aL}{Ak}\ln\frac{h_1}{h_2}$$

or

$$k = 2.303\frac{aL}{At}\lg\frac{h_1}{h_2} \tag{3.19}$$

Example 3.1

Refer to the constant-head permeability test arrangement shown in Figure 3.7. A test gives these values: $L = 30$ cm; $A = 177$ cm^2; $h = 50$ cm; Water collected in a period of 5 min is 350 cm^3.

Calculate the hydraulic conductivity in cm/s.

Solution

From Eq.（3.16），

$$k = \frac{QL}{Aht}$$

Given $Q = 350$ cm^3, $L = 30$ cm, $A = 177$ cm^2, $h = 50$ cm, and $t = 5$ min, we have

$$k = \frac{350 \times 30}{177 \times 50 \times 5 \times 60} = 3.95 \times 10^{-3}\ \text{cm/s}$$

Example 3. 2

For a falling-head permeability test, the following values are given: Length of specimen = 200 mm; Area of soil sample = 1000 mm^2; Area of standpipe = 40 mm^2; Head difference at time t = 0, h_1 = 500 mm; Head difference at time t = 180s, h_2 = 300 mm.

Determine the hydraulic conductivity of the soil in cm/s.

Solution

From Eq. (3. 19),

$$k = 2.303 \frac{aL}{At} \lg\left(\frac{h_1}{h_2}\right)$$

We are given a = 40 mm^2, L = 200 mm, A = 1000 mm^2, t = 180s, h_1 = 500 mm, and h_2 = 300 mm,

$$k = 2.303 \frac{40 \times 200}{1000} \lg\left(\frac{500}{300}\right)$$

$$= 2.27 \times 10^{-2} \text{cm/s}$$

Example 3. 3

A permeable soil layer is underlain by an impervious layer, as shown in Figure 3. 9 (a). With k = 5.3 × 10^{-5} m/s for the permeable layer, calculate the rate of seepage through it in m^3/(h · m) width if H = 3 m and α = 8°.

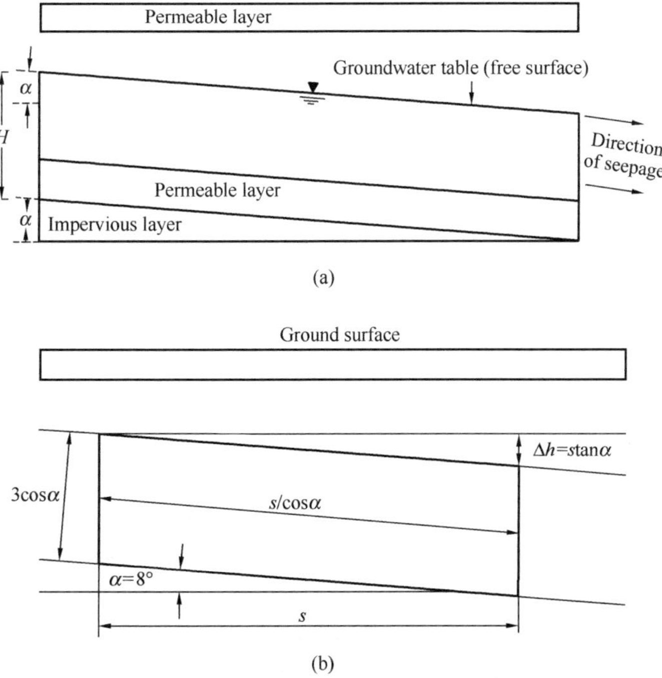

Figure 3. 9 Figure of example 3. 3

Solution

From Figure 3.9 (b),

$$i=\frac{\text{head} \cdot \text{loss}}{\text{length}}=\frac{s\tan\alpha}{\left(\dfrac{s}{\cos\alpha}\right)}=\sin\alpha$$

$$q=kiA=k\times\sin\alpha\times3\cos\alpha\times1$$

$$k=5.3\times10^{-5}\text{m/s}$$

$$q=5.3\times10^{-5}\times\sin8°\times3600\approx0.0265\ \text{m}^3/(\text{h}\cdot\text{m})$$

Example 3.4

Find the flow rate in $\text{m}^3/(\text{s}\cdot\text{m})$ length (at right angles to the cross section shown) through the permeable soil layer shown in Figure 3.10 given $H=8$ m, $H_1=3$ m, $h=$ 4m, $s=50$ m, $\alpha=8°$, and $k=0.08$ cm/s.

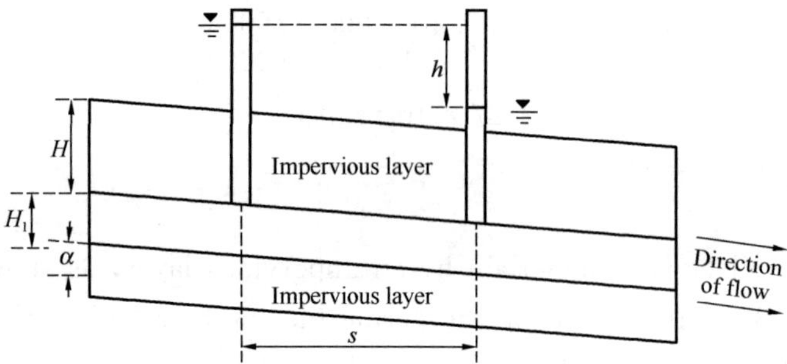

Figure 3.10 Flow through permeable layer

Solution

Hydraulic gradient (i) $= h\cos\alpha/s$

From Eqs. (3.5) and (3.6),

$$q=kiA=k\left(\frac{h\cdot\cos\alpha}{s}\right)(H_1\cos\alpha\times1)$$

$$=(0.08\times10^{-2}\ \text{m/s})\left(\frac{4\cos8°}{50}\right)(3\cos8°\times1)$$

$$=0.19\times10^{-3}\ \text{m}^3/(\text{s}\cdot\text{m})$$

3.5 Field Test Determination of Coefficient of Permeability

In the field, the average hydraulic conductivity of a soil deposit in the direction of flow can be determined by performing pumping tests from wells. Figure 3.11 shows a case where the top permeable layer, whose

3.5 渗透系数现场试验测定

在现场，土的沉积物在渗流方向上的平均渗透系数可以通过从井中进行抽水试验来确定。如图 3.11 所示，顶部渗透层的渗

hydraulic conductivity has to be determined, is unconfined and underlain by an impermeable layer. During the test, water is pumped out at a constant rate from a test well that has a perforated casing. Several observation wells at various radial distances are made around the test well. Continuous observations of the water level in the test well and in the observation, wells are made after the start of pumping, until a steady state is reached. The steady state is established when the water level in the test and observation wells becomes constant. The expression for the rate of flow of groundwater into the well, which is equal to the rate of discharge from pumping, can be written as

透系数是确定的，无约束的，其下方是不透水层。在试验过程中，水以恒定速率从带有穿孔套管的试验井中抽出。在测试井周围设置了几个不同径向距离的观测井。开始抽水后，对试验井和观测井中的水位进行连续观测，直至达到稳定状态。当试验井和观测井中的水位不再变化，说明达到稳定状态。地下水流入井内的速率等于抽水的排放速率，其表达式可写为式（3.20），由此得出式（3.21）。

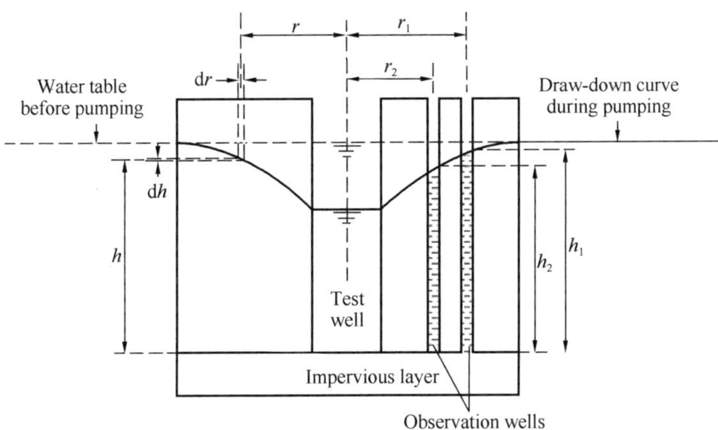

Figure 3.11 Pumping test from a well in an unconfined permeable layer underlain by an impermeable stratum

$$q = k\left(\frac{dh}{dr}\right)2\pi rh \tag{3.20}$$

or

$$\int_{r_2}^{r_1}\frac{dr}{r} = \left(\frac{2\pi k}{q}\right)\int_{h_2}^{h_1} h\,dh$$

Thus,

$$k = \frac{2.303q\lg\left(\frac{r_1}{r_2}\right)}{\pi(h_1^2 - h_2^2)} \tag{3.21}$$

From field measurements, if q, r_1, r_2, h_1, and h_2 are known, the hydraulic conductivity can be calculated from the simple relationship presented in Eq. (3.21). This equation also can be written as

根据现场测量，如果已知 q，r_1，r_2，h_1 和 h_2，则可以根据式（3.21）计算渗透系数。这个式子也可写为式（3.22）。

$$k = \frac{2.303q\lg\left(\frac{r_1}{r_2}\right)}{14.7\pi(h_1^2 - h_2^2)} \tag{3.22}$$

where k is in cm/s, q is in g/m, h_1 and h_2 are in ft.

式中，k 的单位是 cm/s，q 的单位为 g/m，h_1 和 h_2 的单位为英尺。

The average hydraulic conductivity for a confined aquifer can also be determined by conducting a pumping test from a well with a perforated casing that penetrates the full depth of the aquifer and by observing the piezometric level in a number of observation wells at various radial distances (Figure 3.12). Pumping is continued at a uniform rate q until a steady state is reached.

承压含水层的平均渗透系数也可以通过从带有穿孔套管的井中进行抽水试验来确定，该套管穿透整个含水层，并通过在不同径向距离观察多个观测井中的测压管水位来确定（图 3.12）。以均匀的速率 q 继续泵送，直到达到稳定状态。

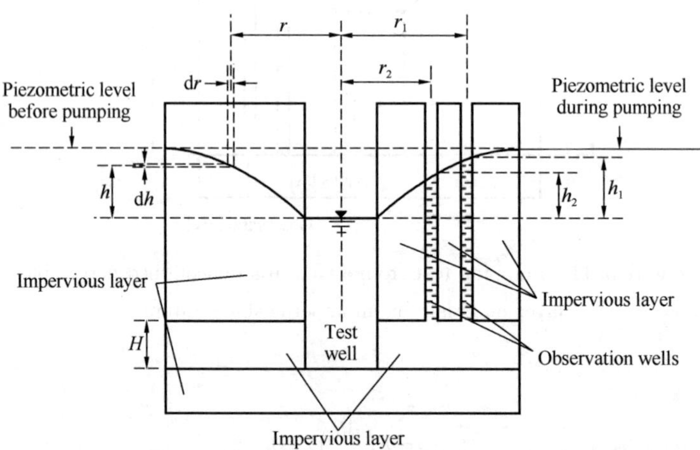

Figure 3.12 Pumping test from a well penetrating the full depth in a confined aquifer

Because water can enter the test well only from the aquifer of thickness H, the steady state of discharge is

由于水只能从厚度为 H 的含水层进入试验井，因此稳定状态的排放量可表达为式（3.23）。

$$q = k\left(\frac{dh}{dr}\right)2\pi rh \tag{3.23}$$

or

$$\int_{r_2}^{r_1} \frac{dr}{r} = \int_{h_2}^{h_1} \frac{2\pi kH}{q} dh$$

This gives the hydraulic conductivity in the direction of flow as

渗流方向上的渗透系数可表达为式（3.24）。

$$k=\frac{q\lg\left(\frac{r_1}{r_2}\right)}{2.727H(h_1-h_2)} \tag{3.24}$$

3.6 2D Flow Problem

In the preceding section, we considered some simple cases for which direct application of Darcy's law was required to calculate the flow of water through soil. In many instances, the flow of water through soil is not in one direction only, nor is it uniform over the entire area perpendicular to the flow. In such cases, the groundwater flow is generally calculated by the use of graphs referred to as flow nets. The concept of the flow net is based on Laplace's equation of continuity, which governs the steady flow condition for a given point in the soil mass.

In this section, we will discuss the following:

(1) Derivation of Laplace's equation of continuity and some simple applications of the equation;

(2) Procedure to construct flow nets and calculation of seepage in isotropic and anisotropic soils;

(3) Seepage through earth dams.

To derive the Laplace differential equation of continuity, let us consider a single row of sheet piles that have been driven into a permeable soil layer, as shown in Figure 3.13 (a). The row of sheet piles is assumed to be impervious. The steady-state flow of water from the upstream to the downstream side through the permeable layer is a two-dimensional flow. For flow at a point A, we consider an elemental soil block. The block has dimensions dx, dy, and dz (length dy is perpendicular to the plane of the paper); it is shown in an enlarged scale in Figure 3.13 (b). Let v_x and v_z be the components of the discharge velocity in the horizontal and vertical directions, respectively. The rate of

3.6 二维渗流问题

上一节讨论了一些直接应用达西定律来计算土中渗流的简单情况。在多数情况下，水在土中的渗流不是单向的，并且在垂直于渗流方向的整个区域也不是均匀的。在这种情况下，地下水流量通常是用流网来计算的。流网的概念基于拉普拉斯连续方程，该方程控制着土中给定点的稳定渗流条件。

本节将讨论以下内容：

（1）拉普拉斯连续方程的推导和简单应用；

（2）在各向同性和各向异性土中构建流网和计算渗流的程序；

（3）土坝渗流。

为推导拉普拉斯连续微分方程，将一排板桩打入透水层，如图 3.13（a）所示。假设板桩是不透水的。水从上游通过渗透层到下游的稳态流动是二维的。对 A 点的流动，构建一个尺寸为 dx、dy 和 dz（长度 dy 垂直于纸面）的基本单元，如图 3.13（b）所示。设 v_x 和 v_z 分别为水平和垂直方向上的排放速度分量，水在水平方向上流入土体单元的速率为 v_xdzdy，在垂直方向上为 v_zdxdy。那么土体单元

flow of water into the elemental block in the horizontal direction is equal to $v_x\,dzdy$, and in the vertical direction it is $v_z\,dxdy$. The rates of outflow from the block in the horizontal and vertical directions are, respectively,

在水平和垂直方向上的渗流速率分别为式（3.25）和式（3.26）。

$$\left(v_x + \frac{\partial v_x}{\partial x}dx\right)dzdy \tag{3.25}$$

and

$$\left(v_z + \frac{\partial v_z}{\partial z}dz\right)dxdy \tag{3.26}$$

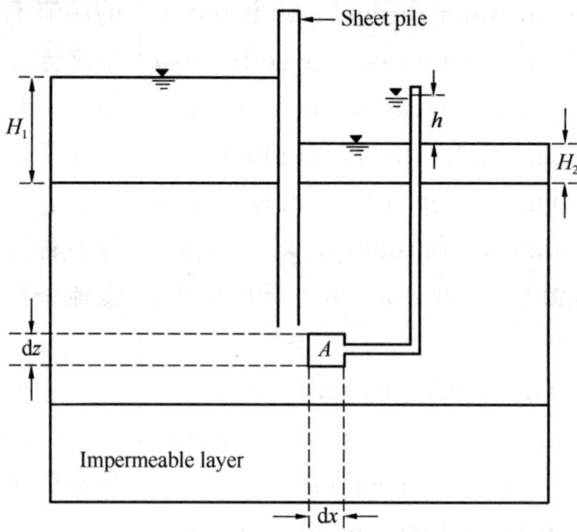

(a) Single-row sheet piles driven into permeable layer

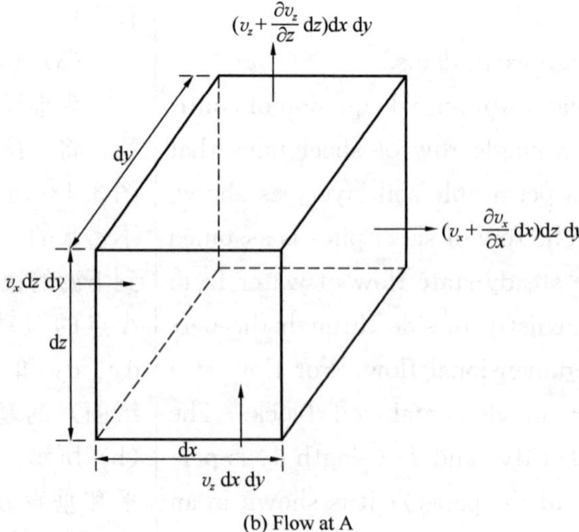

(b) Flow at A

Figure 3.13 Single-row sheet piles driven into permeable layer and flow at A

Assuming that water is incompressible and that no volume change in the soil mass occurs, we know that the total rate of inflow should equal the total rate of outflow. Thus,

假设水是不可压缩的，土中没有体积变化，那么总流入率应该等于总流出率，即式（3.27）或式（3.28）。

$$\left[\left(v_x+\frac{\partial v_x}{\partial x}\mathrm{d}x\right)\mathrm{d}z\mathrm{d}y+\left(v_z+\frac{\partial v_z}{\partial z}\mathrm{d}z\right)\mathrm{d}x\mathrm{d}y\right]-\left[v_x\mathrm{d}z\mathrm{d}y+v_z\mathrm{d}x\mathrm{d}y\right]=0 \quad (3.27)$$

or

$$\frac{\partial v_x}{\partial x}+\frac{\partial v_z}{\partial z}=0 \quad (3.28)$$

With Darcy's law, the discharge velocities can be expressed as

根据达西定律，排放速度可表示为式（3.29）和式（3.30）。

$$v_x=k_x i_x=k_x\frac{\partial h}{\partial x} \quad (3.29)$$

and

$$v_z=k_z i_z=k_z\frac{\partial h}{\partial z} \quad (3.30)$$

where k_x and k_z are the hydraulic conductivities in the horizontal and vertical directions, respectively.

From Eqs. (3.28), (3.29), and (3.30), we can write

式中，k_x和k_z分别为水平和垂直方向上的渗透系数。根据式（3.28）、式（3.29）和式（3.30），可得式（3.31）。

$$k_x\frac{\partial^2 h}{\partial x^2}+k_z\frac{\partial^2 h}{\partial z^2}=0 \quad (3.31)$$

If the soil is isotropic with respect to the hydraulic conductivity, that is $k_x=k_z$, the preceding continuity equation for two-dimensional flow simplifies to

如果土的渗透系数是各向同性的，即$k_x=k_z$，则式（3.31）可以简化为式（3.32）。

$$\frac{\partial^2 h}{\partial x^2}+\frac{\partial^2 h}{\partial z^2}=0 \quad (3.32)$$

3.7 Continuity Equation for Solution of Simple 2D Flow Problems

3.7 简单的二维渗流计算

The continuity equation given in Eq. (3.32) can be used in solving some simple flow problems. To illustrate this, let us consider a one-dimensional flow problem, as shown in Figure 3.14, in which a constant head is maintained across a two-layered soil for the flow of water. The head difference between the top of soil layer

式（3.32）中给出的连续方程可用于解决一些简单的渗流问题。为更好地说明，将其看为一维渗流问题，如图 3.14 所示，在两层土中保持恒定的水头以维持水流。第一层土顶部和第二层

No. 1 and the bottom of soil layer No. 2 is h_1.

土底部之间的水头差为 h_1。

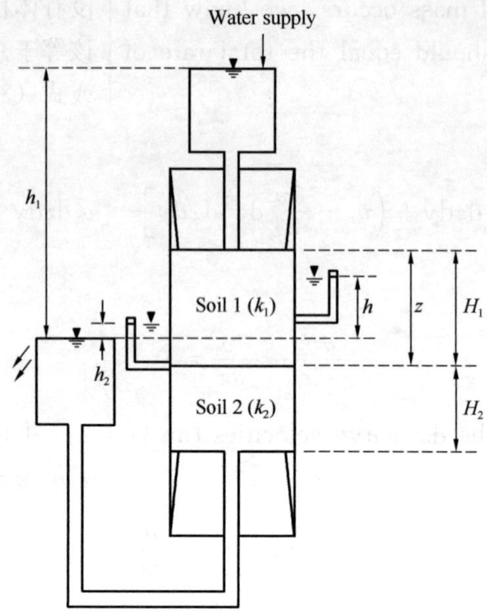

Figure 3.14 Flow through a two-layered soil

Because the flow is in only the z direction, the Eq. (3.32) is simplified to the form:

由于仅在 z 方向上渗流，因此连续方程式（3.32）被简化为式（3.33）或式（3.34）。

$$\frac{\partial^2 h}{\partial z^2} = 0 \tag{3.33}$$

or

$$h = A_1 z + A_2 \tag{3.34}$$

where A_1 and A_2 are constants.

式中，A_1 和 A_2 为常数。

To obtain A_1 and A_2 for flow through soil layer No. 1, we must know the boundary conditions, which are as follows:

Condition 1: At $z=0$, $h=h_1$.

Condition 2: At $z=H_1$, $h=h_2$.

Combining Eq. (3.34) and Condition 1 gives

为了得出第一层土的 A_1 和 A_2，必须确定边界条件

条件 1：$z=0$，$h=h_1$；

条件 2：$z=H_1$，$h=h_2$，将条件 1 代入式（3.34）可得式（3.35）；

$$A_2 = h_1 \tag{3.35}$$

Similarly, combining Eq. (3.34) and Condition 2 with Eq. (3.35) gives

同样，将条件 2 代入式（3.34），结合式（3.35），得式（3.36）或式（3.37）。

$$h_2 = A_1 H_1 + h_1 \tag{3.36}$$

or

$$A_1 = -\left(\frac{h_1 - h_2}{H_1}\right) \tag{3.37}$$

Combining Eqs. (3.34), (3.35), and (3.37),

结合式（3.34）、式（3.35）

we obtain

和式（3.37），可得式（3.38）。

$$h = -\left(\frac{h_1 - h_2}{H_1}\right)z + h_1 \qquad (\text{for } 0 \leqslant z \leqslant H_1) \tag{3.38}$$

For flow through soil layer No. 2, the boundary conditions are

Condition 1: At $z = H_1$, $h = h_2$.

Condition 2: At $z = H_1 + H_2$, $h = 0$.

From Condition 1 and Eq. (3.34),

第二层土的边界条件为条件 1：$z = H_1$，$h = h_2$；条件 2：$z = H_1 + H_2$，$h = 0$。将条件 1 代入式（3.34）得式（3.39），

$$A_2 = h_2 - A_1 H_1 \tag{3.39}$$

Also, from Condition 2 and Eqs. (3.34) and (3.39),

将条件 2 代入式（3.34），结合式（3.39），得式（3.40）。

$$0 = A_1(H_1 + H_2) + (h_2 - A_1 H_1)$$
$$A_1 H_1 + A_1 H_2 + h_2 - A_1 H_1 = 0$$

or

$$A_1 = -\frac{h_2}{H_2} \tag{3.40}$$

So, from Eqs. (3.34), (3.39), and (3.40),

由式（3.34）、式（3.39）和式（3.40）得式（3.41）。

$$h = -\left(\frac{h_2}{H_2}\right)z + h_2\left(1 + \frac{H_1}{H_2}\right) \qquad (\text{for } H_1 \leqslant z \leqslant H_1 + H_2) \tag{3.41}$$

At any given time, flow through soil layer No. 1 equals flow through soil layer No. 2, so

在任何给定时间，流经第一层土的流量等于流经第二层土的流量，因此有式（3.42）或式（3.43）。

$$q = k_1\left(\frac{h_1 - h_2}{H_1}\right)A = k_2\left(\frac{h_2 - 0}{H_2}\right)A \tag{3.42}$$

where A = area of cross section of the soil;

k_1 = hydraulic conductivity of soil layer No. 1;

k_2 = hydraulic conductivity of soil layer No. 2

式中，A 为土的横截面面积，k_1 为第一层土的渗透系数，k_2 为第二层土的渗透系数。

or

$$h_2 = \frac{h_1 k_1}{H_1\left(\frac{k_1}{H_1} + \frac{k_2}{H_2}\right)} \tag{3.43}$$

Substituting Eq. (3.43) into Eq. (3.38), we obtain

将式（3.43）代入式（3.38）中，得式（3.44），

$$h = h_1\left(1 - \frac{k_2 z}{k_1 H_2 + k_2 H_1}\right) \qquad (\text{for } 0 \leqslant z \leqslant H_1) \tag{3.44}$$

Similarly, combining Eqs. (3.41) and (3.43) gives

结合式（3.41）和式（3.43），得式（3.45）。

$$h = h_1\left[\left(\frac{k_1}{k_1 H_2 + k_2 H_1}\right)(H_1 + H_2 - z)\right] \quad (\text{for } H_1 \leqslant z \leqslant H_1 + H_2) \tag{3.45}$$

Example 3.5

Refer to Figure 3.14. Given: $H_1 = 305$ mm, $H_2 = 508$ mm, $h_1 = 610$ mm, $h = 508$ mm, $z = 203$ mm, $k_1 = 0.066$ cm/s, and diameter of the soil sample is $D = 76$ mm.

Determine the rate of flow of water through the two-layered soil (cm^3/h).

例题 3.5

如图 3.14 所示，当 $H_1 = 305$ mm，$H_2 = 508$ mm，$h_1 = 610$ mm，$h = 508$ mm，$z = 203$ mm，$k_1 = 0.066$cm/s，土样的直径 $D = 76$ mm，求过两层土的渗透速度（cm^3/h）。

Solution

Since $z = 203$ mm is located in soil layer 1, Eq. (3.44) is valid. Thus,

解得 $q \approx 5390\ \text{cm}^3/\text{h}$。具体过程见英文。

$$h = h_1\left(1 - \frac{k_2 z}{k_1 H_2 + k_2 H_1}\right) = h_1\left[1 - \frac{z}{\left(\frac{k_1}{k_2}\right) H_2 + H_1}\right]$$

$$508 = 610\left[1 - \frac{203}{\left(\frac{k_1}{k_2}\right) 508 + 305}\right]$$

$$\frac{k_1}{k_2} = 1.795 \approx 1.8$$

Given $k_1 = 0.066$ cm/s. Thus,

$$k_2 = \frac{k_1}{1.8} = \frac{0.066}{1.8} \approx 0.037\text{cm/s}$$

This rate of flow is

$$q = k_{\text{eq}} i A$$

$$i = \frac{h_1}{H_1 + H_2} = \frac{610}{305 + 508} \approx 0.75$$

$$A = \frac{\pi}{4} D^2 = \frac{\pi}{4}(7.6)^2 \approx 45.36\text{cm}^2$$

$$k_{\text{eq}} = \frac{H_1 + H_2}{\frac{H_1}{k_1} + \frac{H_2}{k_2}} \approx \frac{30.5 + 50.8}{\frac{30.5}{0.066} + \frac{50.8}{0.037}} \approx 0.0443\text{cm/s} = 159.48\text{cm/h}$$

Thus,

$$q = k_{\text{eq}} i A = 159.48 \times 0.75 \times 45.06 \approx 5390\text{cm}^3/\text{h}$$

3.8 Flow Nets

The Eq. (3.39) in an isotropic medium represents

3.8 流网

各向同性介质中的连续方程

two orthogonal families of curves, that is, the flow lines and the equipotential lines. A flow line is a line along which a water particle will travel from upstream to the downstream side in the permeable soil medium. An equipotential line is a line along which the potential head at all points is equal. Thus, if piezometers are placed at different points along an equipotential line, the water level will rise to the same elevation in all of them. Figure 3. 15 (a) demonstrates the definition of flow and equipotential lines for flow in the permeable soil layer around the row of sheet piles shown in Figure 3. 13 (for $k_x = k_z = k$) .

式 (3. 39) 表示两个正交的曲线族，即流线和等势线。流线是水质点在可渗透土介质中从上游向下游行进的路线。等势线是所有势能相等的点的连线。因此，如果压力计沿着等势线放置在不同的点上，所有这些点的水位都会上升到相同的高度。图 3. 15 (a) 解释了图 3. 13 板桩周围渗透土层中流线和等势线的定义 ($k_x = k_z = k$)。

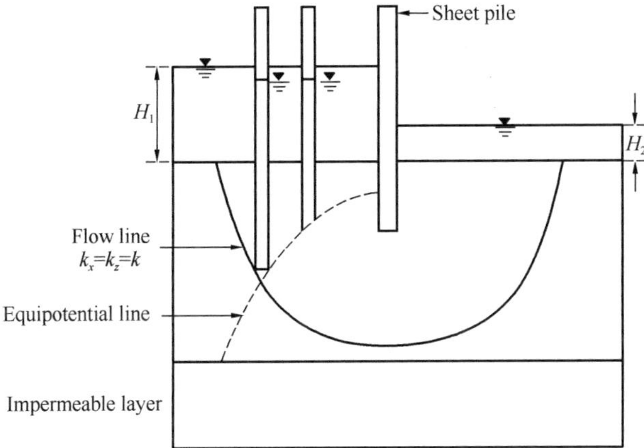

(a) Definition of flow lines and equipotential lines

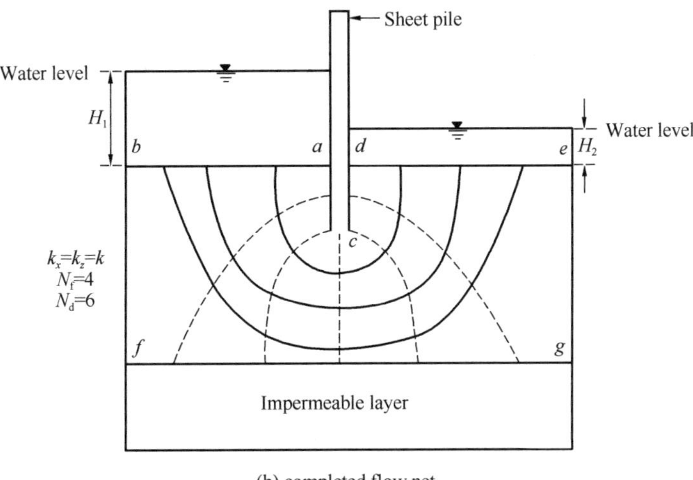

(b) completed flow net

Figure 3. 15 Flow Lines, equipotential lines, and flow net around sheet piles

A combination of a number of flow lines and equipotential lines is called a flow net. Flow nets are constructed for the calculation of groundwater flow and the evaluation of heads in the media. To complete the graphic construction of a flow net, one must draw the flow and equipotential lines in such a way that:

1) The equipotential lines intersect the flow lines at right angles.

2) The flow elements formed are approximate squares.

Figure 3.15 (b) shows an example of a completed flow net. One more example of flow net in isotropic permeable layer is given in Figure 3.16. In these figures, N_f is the number of flow channels in the flow net, and N_d is the number of potential drops.

多条流线和等势线的组合称为流网，用来计算介质中的地下水流量和水头。为完成流网的图形构建，必须以以下方式绘制流网和等势线：

1）等势线与流线垂直相交。

2）形成的流动单元近似正方形。

图 3.15（b）为一个完整的流网示例。图 3.16 为各向同性渗透层中流网的另一个例子。在这些图中，N_f是流网中的流槽数，N_d是势降数。

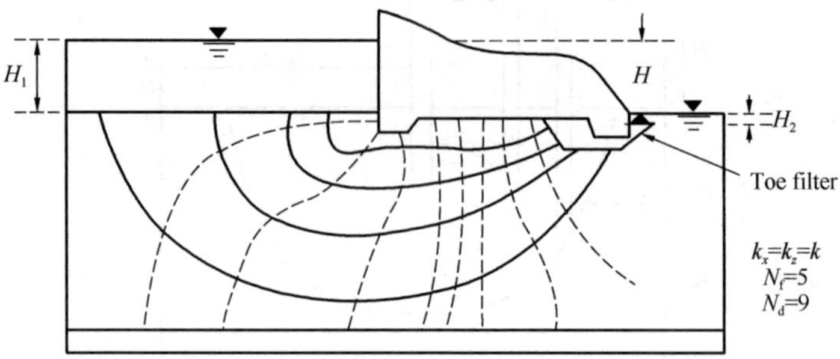

Figure 3.16 Flow net under a dam with toe filter

Drawing a flow net takes several trials. While constructing the flow net, keep the boundary conditions in mind. For the flow net shown in Figure3.15 (b), the following four boundary conditions apply:

1) The upstream and downstream surfaces of the permeable layer (lines *ab* and *de*) are equipotential lines.

2) Because *ab* and *de* are equipotential lines, all the flow lines intersect them at right angles.

3) The boundary of the impervious layer, that is, line *fg* is a flow line, and so is the surface of the impervious sheet pile, line *acd*.

4) The equipotential lines intersect acd and *fg* at right angles.

绘制流网需要进行多次尝试，同时在构建流网时，边界条件也是必要的。图 3.15（b）所示流网四个边界条件如下：

1）渗透层的上游和下游表面（线 *ab* 和 *de*）是等势线。

2）因为 *ab* 和 *de* 是等势线，所以所有流线都与它们垂直相交。

3）不透水层的边界（即线 *fg*）是一条流线，板桩的表面也是一条流线，即线 *acd*。

4）等势线与 *acd* 和 *fg* 垂直相交。

3.9 Seepage Calculation From a Flow Net

In any flow net, the strip between any two adiacent flow lines is called a flow channel. Figure 3.17 shows a flow channel with the equipotential lines forming square elements. Let h_1, h_2, h_3, h_4,..., h_n be the piezometric levels corresponding to the equipotential lines. The rate of seepage through the flow channel per unit length (perpendicular to the vertical section through the permeable layer) can be calculated as follows. Because there is no flow across the flow lines.

3.9 流网的渗流计算

在流网中，任意两条流线之间的条带称为流槽。图 3.17 为一等势线形成的正方形单元的流槽。设 h_1，h_2，h_3，h_4，…，h_n是与等势线对应的测压管水头。单位长度内（垂直于渗透层的垂直截面）通过流槽的渗流率为式(3.46)。

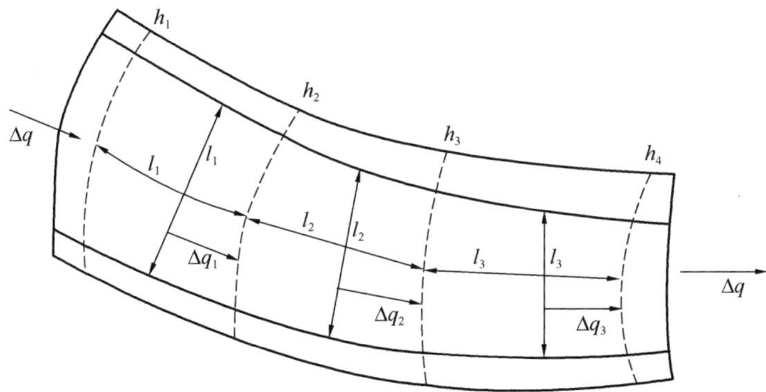

Figure 3.17 Seepage through a flow channel with square elements

$$\Delta q_1 = \Delta q_2 = \Delta q_3 = \cdots = \Delta q \tag{3.46}$$

From Darcy's law, the flow rate is equal to kiA. Thus, Eq. (3.46) can be written as

因为流线上没有水流通过，根据达西定律，渗流率等于 kiA，因此式（3.46）可以写为式（3.47）。

$$\Delta q = k\left(\frac{h_1 - h_2}{l_1}\right)l_1 = k\left(\frac{h_2 - h_3}{l_2}\right)l_2 = k\left(\frac{h_3 - h_4}{l_3}\right)l_3 = \cdots \tag{3.47}$$

Equation (3.47) shows that if the flow elements are drawn as approximate squares, the drop in the piezometric level between any two adjacent equipotential lines is the same. This is called the potential drop. Thus,

式（3.47）表明，如果将水流单元绘制为近似正方形，则任何两条相邻等势线之间的测压管水位下降都是相同的，称为势降。因此，有式（3.48）和式(3.49)成立。

$$h_1 - h_2 = h_2 - h_3 = h_3 - h_4 = \cdots = \frac{H}{N_d} \tag{3.48}$$

and

$$\Delta q = k\frac{H}{N_{\mathrm{d}}} \tag{3.49}$$

where H = head difference between the upstream and downstream sides

N_{d} = number of potential drops

In Figure 3.15 (b), for any flow channel, $H = H_1 - H_2$ and $N_{\mathrm{d}} = 6$.

式中，H为上游和下游侧之间的水头差，N_{d}为势降数。

在图3.15（b）中，对于任何流槽，$H = H_1 - H_2$，$N_{\mathrm{d}} = 6$。

If the number of flow channels in a flow net is equal to N, the total rate of flow through all the channels per unit length can be given by

如果流网中的流槽数为N，那么单位长度通过所有流槽的总流量可由式（3.50）计算出。

$$q = k\frac{HN_{\mathrm{f}}}{N_{\mathrm{d}}} \tag{3.50}$$

Although drawing square elements for a flow net is convenient, it is not always necessary. Alternatively, one can draw a rectangular mesh for a flow channel, as shown in Figure 3.18, provided that the width-to-length ratios for all the rectangular elements in the flow net are the same. In this case, Eq. (3.47) for rate of flow through the channel can be modified to

虽然绘制流槽的正方形单元很方便，但不是必要的，可以在流槽中绘制矩形网格，如图3.18所示，前提是流槽中所有矩形单元的宽长比相同。此时，通过流槽的流速，即式（3.47）可写为式（3.51）。

$$\Delta q = k\left(\frac{h_1 - h_2}{l_1}\right)b_1 = k\left(\frac{h_2 - h_3}{l_2}\right)b_2 = k\left(\frac{h_3 - h_4}{l_3}\right)b_3 = \cdots \tag{3.51}$$

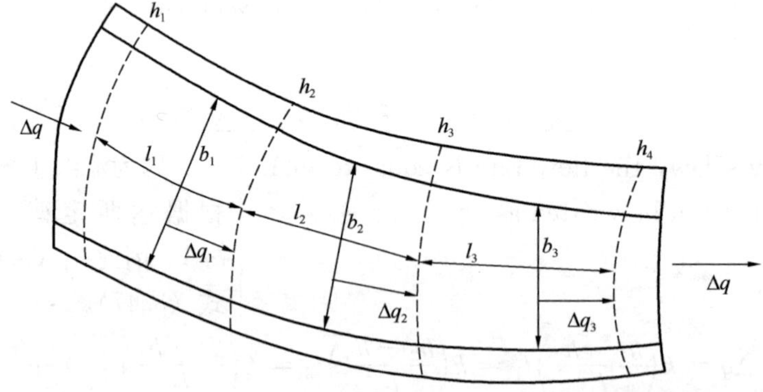

Figure 3.18 Seepage through a flow channel with rectangular elements

If $b_1/l_1 = b_2/l_2 = b_3/l_3 = \cdots = n$ (i.e., the elements are not square), Eqs. (3.49) and (3.50) can be modified to

如果$b_1/l_1 = b_2/l_2 = b_3/l_3 = \cdots = n$（即不是正方形单元），则式（3.49）和式（3.50）可以改为式（3.52）和式（3.53）。

$$\Delta q = kH\left(\frac{n}{N_{\mathrm{d}}}\right) \tag{3.52}$$

and

$$q = kH\left(\frac{N_f}{N_d}\right)n \tag{3.53}$$

Figure 3.19 shows a flow net for seepage around a single row of sheet piles. Note that flow channels 1 and 2 have square elements. Hence, the rate of flow through these two channels can be obtained from Eq. (3.49):

图 3.19 为单排板桩周围渗流的流网，流槽 1 和流槽 2 为正方形单元。因此，通过这两个流槽的流速可以根据式（3.49）计算得出：

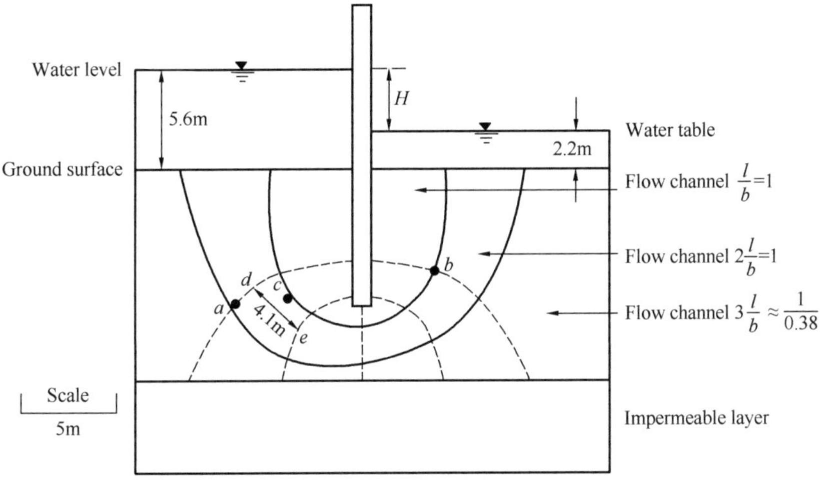

Figure 3.19 Flow net for seepage around a single row of sheet piles

$$\Delta q_1 + \Delta q_2 = \frac{k}{N_d}H + \frac{k}{N_d}H = \frac{2kH}{N_d}$$

However, flow channel 3 has rectangular elements. These elements have a width-to-length ratio of about 0.38; hence, from Eq. (3.52),

然而，流槽 3 为矩形单元，宽长比约为 0.38，通过此流槽的流速可以根据式（3.52）得出，

$$\Delta q_3 = \frac{k}{N_d}H(0.38)$$

So, the total rate of seepage can be given as

因此，总渗透速率可以表示为式（3.54）。

$$q = \Delta q_1 + \Delta q_2 + \Delta q_3 = 2.38\frac{kH}{N_d} \tag{3.54}$$

Example 3.6

A flow net for flow around a single row of sheet piles in a permeable soil layer is shown in Figure 3.19. Given that $k_x = k_z = k = 5 \times 10^{-3}$ cm/s, determine

a. How high (above the ground surface) the water will rise if piezometers are placed at points a and b.

b. The total rate of seepage through the permeable layer per unit length.

c. The approximate average hydraulic gradient at *c*.

Solution

Part a

From Figure 3.19, we have $N_d = 6$, $H_1 = 5.6\text{m}$, and $H_2 = 2.2\text{m}$. So, the head loss of each potential drop is

$$\Delta H = \frac{H_1 - H_2}{N_d} = \frac{5.6 - 2.2}{6} = 0.567\text{m}$$

At point *a*, we have gone through one potential drop. So, the water in the piezometer will rise to an elevation of

$$(5.6 - 0.567) = 5.033\text{m (above the ground surface)}$$

At point *b*, we have five potential drops. So, the water in the piezometer will rise to an elevation of

$$5.6 - 5 \times 0.567 = 2.765\text{m (above the ground surface)}$$

Part b

From Eq. (3.54),

$$q = 2.38\frac{k(H_1 - H_2)}{N_d} = \frac{2.38 \times 5 \times 10^{-5} \times (5.6 - 2.2)}{6}$$
$$= 6.74 \times 10^{-5}\text{m}^3/(\text{s} \cdot \text{m})$$

Part c

The average hydraulic gradient at *c* can be given as:

$$i = \frac{\Delta H}{\Delta L} = \frac{0.567\text{m}}{4.1\text{m}} \approx 0.138$$

(*Note*: The average length of flow has been scaled.)

Problems

3.1 A permeable soil layer is underlain by an impervious layer as shown in Figure 3.20. Knowing that $k = 4.8 \times 10^{-3}$ cm/s for the permeable layer, calculate the rate of seepage through this layer in $\text{m}^3/(\text{h} \cdot \text{m})$ width. Given: $H = 4.2\text{m}$ and $\alpha = 6°$.

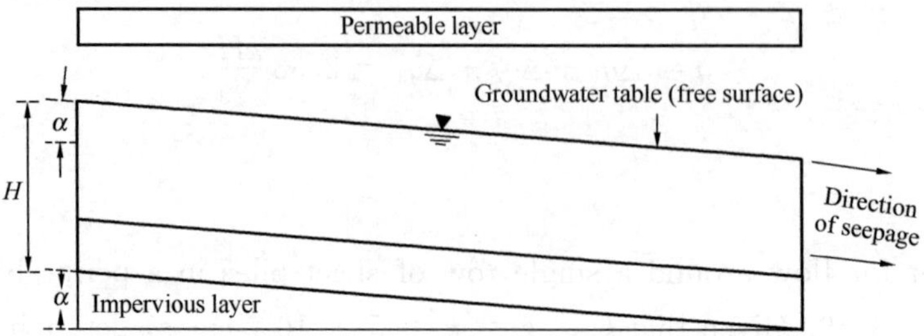

Figure 3.20 Figure of problem 3.1

3.2 Find the rate of flow in $m^3/(s \cdot m)$ (at right angles to the cross section shown in Figure 3.21) through the permeable soil layer. Given: $H = 4m$, $H_1 = 2m$, $h = 2.75m$, $S = 30m$, $\alpha = 14°$, and $k = 0.075cm/s$.

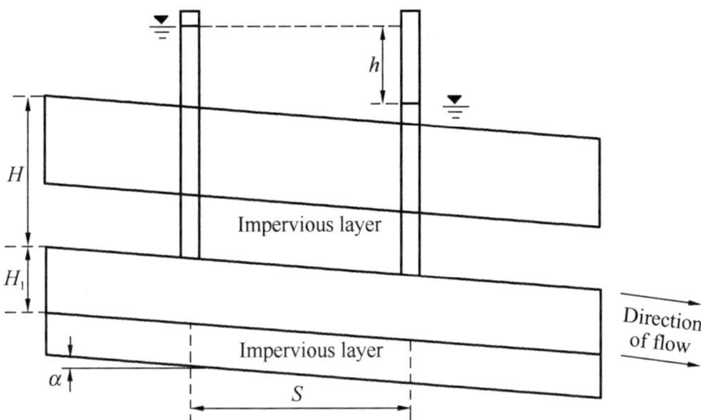

Figure 3.21 Figure of problem 3.2

3.3 The results of a constant-head permeability test for a fine sand sample having a diameter of 150 mm and a length of 300 mm are as follows (refer to Figure 3.7):

Constant-head difference = 500mm;

Water collected in 5min = $350cm^3$;

Void ratio of sand = 0.61.

Determine:

a. Hydraulic conductivity, k (cm/s);

b. Seepage velocity.

3.4 The following data are for a falling-head permeability test:

Length of the soil sample = 150mm;

Area of the soil sample = $1964mm^2$;

Area of the standpipe = $25mm^2$.

At time $t = 0$, head difference = 400mm;

At time $t = 8min$, head difference = 200mm.

a. Determine the hydraulic conductivity of the soil (cm/s) .

b. What was the head difference at $t = 6min$?

3.5 Refer to the constant-head permeability test arrangement in a two-layered soil as shown in Figure 3.22. During the test, it was seen that when a constant head of $h_1 = 200mm$ was maintained, the magnitude of h_2 was 80mm. If k_1 is 0.004cm/s, determine the value of k_2 given $H_1 = 100mm$ and $H_2 = 150mm$.

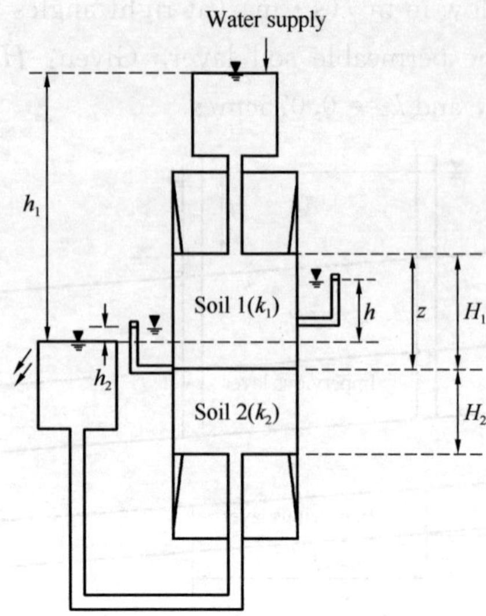

Figure 3. 22　Figure of problem 3. 5

3. 6　A homogeneous anisotropic embankment dam section is detailed in Figure 3. 23, the coefficient permeability in the x and z directions being 4.5×10^{-8} and 1.6×10^{-8} m/s, respectively. Construct the flow net and determine the quantity of seepage through the dam. What is the pore water pressure at point P?

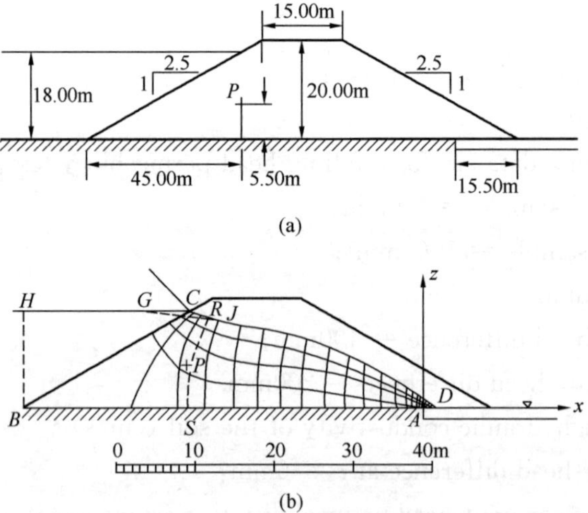

Figure 3. 23　Figure of problem 3. 6

Chapter 4 Stress in Soil

4.1 Introduction

When the load of the structure acts on the ground through the foundation, it will induce stress and deformation in subsoil. This will result in two engineering problems related to the structure: the stability (strength) of soil and the deformation of soil.

On the one hand, if the induced shear stress in subsoil is within the allowable range of soil strength, then the soil is stable; On the contrary, damage will occur. On the other hand, if the deformation of the subsoil exceeds the allowable value, the structure may also be damaged and lose its functional use. Therefore, in order to ensure the safety and normal use of the structure, it is necessary to study the subsoil stress state and distribution law under various load conditions. The calculation and distribution pattern of stress will be discussed in this chapter.

Soil can withstand both the load of the structure and its own weight, which generates subsoil stress. Generally speaking, the stress in subsoil can be divided into two categories. Self-weight stress is stress generated by the weight of soil before the construction of a building. Additional stress refers to the stress caused by external loads such as the weight of the building on the ground after its construction. " Additional" refers to the pressure added on the basis of self-weight stress. In addition, the calculation method of foundation base pressure and the principle of effective stress will also be introduced.

Overall, in this chapter, we will

(1) Determine the vertical stress caused by soil self-weight and foundation base pressure;

第 4 章 土中应力

4.1 概述

当结构的荷载通过基础作用于地面时，将在地基中引起应力和变形。这就会产生两个与结构相关的工程问题，即土的稳定性（强度）问题和土的变形问题。

一方面，如果在地基中诱发的剪应力在土体强度的允许范围内，则土是稳定的；反之，则会发生破坏。另一方面，如果地基的变形超过允许值，结构也可能受到破坏，失去使用功能。因此，为了保证结构的安全和正常使用功能，必须对各种荷载条件下的地基应力状态和地基应力分布规律进行研究。本章将讨论应力的计算及其分布模式。

土体既能承受结构的荷载，也能承受其自重，这些荷载会引起地基应力。一般来说，地基土中的应力可分为两类：自重应力是在建筑物建成之前由土体重量产生的应力；附加应力是建筑物建成后，由建筑物在地面上的重量等外部荷载引起的应力，“附加”是指在自重应力的基础上增加的应力。此外，还将介绍基底压力的计算方法和有效应力原理。

总的来说，在本章中

（1）将确定由土体自重和基底压力引起的竖向应力；

（2） Understand the relationship between total stress, pore water pressure, and effective stress, as well as the importance of effective stress in soil mechanics.

（2）了解总应力、孔隙水压力和有效应力之间的关系，以及有效应力在土力学中的重要性。

4.2 Self-weight Stress

4.2 自重应力

To confirm the initial state of stress in soil, we need to know the self-weight stress. Self-weight stress is the stress generated by the weight of soil before the construction of a building. Now we will discuss how to calculate self-weight stress: assuming that the ground is a semi-infinite, uniform, linearly, isotropic elastic material. As shown in Figure 4.1 (a), consider the soil elements at a given depth of z below the surface. If the unit weight of soil is γ (kN/m^3), the vertical stress can be calculated by calculating the weight of the vertical soil column. At a depth of z, a unit horizontal area is taken as the support, and the weight of the column is divided by its bottom area to obtain

为确定土中应力的初始状态，需要知道自重应力。自重应力是指建筑物建造前土体重量产生的应力，现在讨论如何计算自重应力：假设地基是半无限、均匀、线性、各向同性的弹性材料。如图 4.1(a) 所示，考虑地表以下给定深度 z 处的土体单元。如果土的重度为 γ（kN/m^3），则可以通过计算垂直土柱的重量来计算竖向应力。在 z 深度处，以单位水平面积作为支撑，用土柱的重量除以其底部面积，可得式（4.1），

$$\sigma_{cz} = \gamma z \tag{4.1}$$

where σ_{cz} is the vertical self-weight stress.

式中，σ_{cz} 是竖向自重应力。

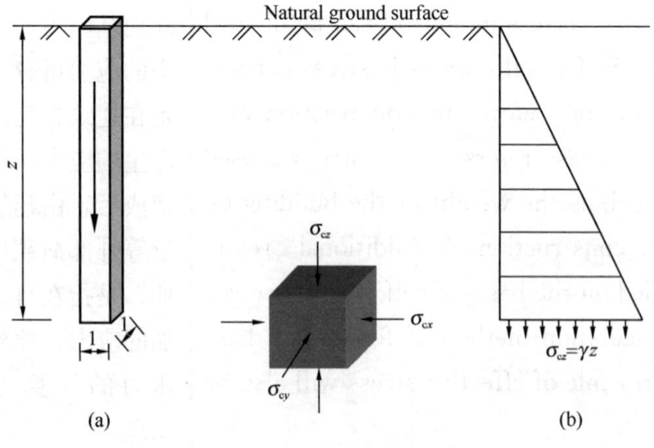

Figure 4.1 Vertical self-weight stress in homogeneous soil

According to the Eq. (4.1), the vertical stress increases linearly with depth and exhibits a triangular distribution, as shown in Figure 4.1(b) . Due to the static

根据式（4.1），竖向应力随深度增加呈线性增加，并呈三角形分布，如图 4.1(b) 所示。由

symmetry of the soil in half space, it is impossible for lateral deformation to occur. Therefore, the two equal horizontal stresses σ_{cx} and σ_{cy} of the soil element can be expressed as:

于土在半空间中的静态对称性，不可能发生横向变形。因此，土体单元的两个相等的水平应力 σ_{cx} 和 σ_{cy} 可以表示为式 (4.2)，

$$\sigma_{cx} = \sigma_{cy} = K_0\sigma_{cz} = K_0\gamma z \tag{4.2}$$

where K_0 is the lateral pressure coefficient of the soil. And the shear stress in soil is 0.

式中，K_0 是土的侧向压力系数，土中的剪应力为 0。

If the calculation point is below the groundwater level, the buoyant unit weight γ' should be used to calculate the self-weight of the underwater part of the column.

如果计算点低于地下水位，则应使用浮重度 γ' 来计算土柱水下部分的自重。

If the ground is composed of multiple layers of subsoil, where the thickness and unit weight of the soil layers are h_1, h_2,…,h_i,…,h_n and γ_1, γ_2,…,γ_i,…,γ_n respectively, it can be obtained that the vertical stress is:

如果地基由多层土组成，其中土层的厚度和重度分别为 h_1，h_2,…,h_i,…,h_n 和 γ_1，γ_2,…,γ_i,…,γ_n，则可以得出竖向应力为式 (4.3)

$$\sigma_{cz} = \sum_{i=1}^{n}\gamma_i h_i \tag{4.3}$$

The horizontal stress is:

水平应力为式 (4.4)。

$$\sigma_{cx} = \sigma_{cy} = K_0\sigma_{cz} = K_0\sum_{i=1}^{n}\gamma_i h_i \tag{4.4}$$

where n is the number of soil layers, h_i is the thickness of the i-th soil layer, γ_i is the natural unit weight of the i-th soil layer.

式中，n 是土层数；h_i 是第 i 层土层的厚度；γ_i 是第 i 个土层的天然重度。

The vertical self-weight stress distribution of multi-layered soil is shown in Figure 4.2.

多层土的竖向自重应力分布如图 4.2 所示。

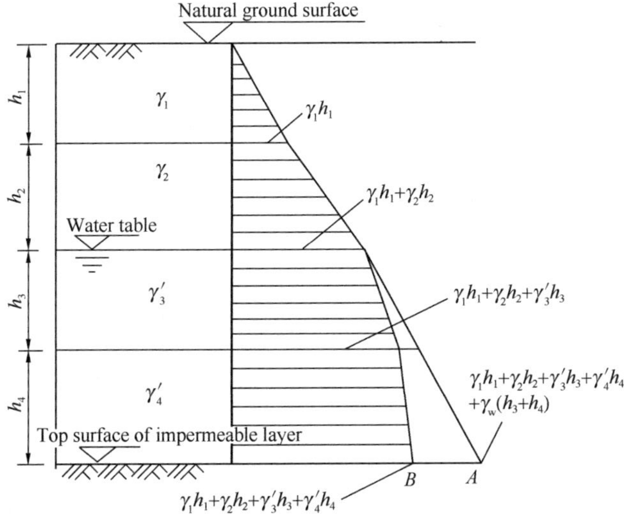

Figure 4.2 Vertical self-weight stress distribution in multi-layered soil

Example 4.1

A layer of saturated clay with a thickness of 4m is covered with 5m deep sand, and the groundwater level is 3m, as shown in Figure 4.3. The saturated unit weight of clay and sand is $19kN/m^3$ and $20\ kN/m^3$, respectively; The unit weight of sand above the groundwater level is $17kN/m^3$. Draw the value of vertical self-weight stress with depth.

例题 4.1

如图 4.3 所示，一层 4m 厚的饱和黏土上覆盖着 5m 厚的砂，地下水位为 3m。黏土和砂的饱和重度分别为 $19kN/m^3$ 和 $20kN/m^3$；地下水位以上砂的重度为 $17kN/m^3$。绘制竖向自重应力随深度的变化曲线。

Solution

解答

$h_1 = 3m, \gamma_1 = 17kN/m^3$; $h_2 = 2m, \gamma'_2 = \gamma_{sat} - \gamma_w = 20 - 10 = 10kN/m^3$; $h_3 = 4m$, $\gamma'_3 = \gamma_{sat} - \gamma_w = 19 - 10 = 9kN/m^3$

$$\sigma_{cz1} = \gamma_1 h_1 = 17 \times 3 = 51kPa$$

$$\sigma_{cz2} = \gamma_1 h_1 + \gamma'_2 h_2 = 51 + 10 \times 2 = 71kPa$$

$$\sigma_{cz3} = \gamma_1 h_1 + \gamma'_2 h_2 + \gamma'_3 h_3 = 71 + 9 \times 4 = 107kPa$$

The distribution of vertical self-weight stress along depth is shown in Figure 4.3.

竖向自重应力沿深度的分布如图 4.3 所示。

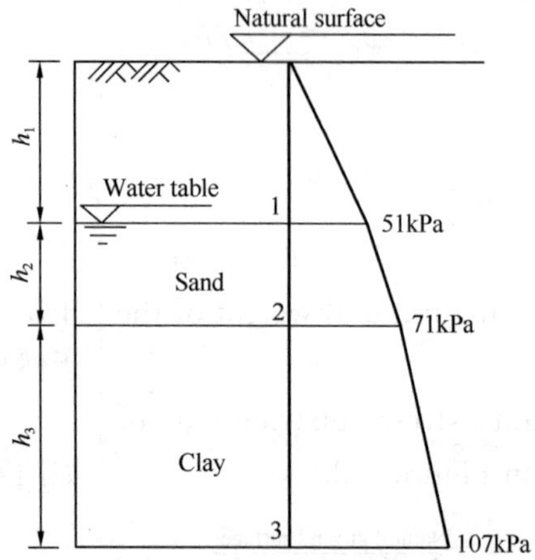

Figure 4.3 Distribution of vertical self-weight stress

4.3 Contact Pressure of Foundation Base

Basically, every structure is supported by soil and rock. The main function of the foundation is to transfer the load of the structure to the supporting ground. Therefore, analyzing the interaction between the foundation and the supporting soil is of great significance for the safety of the structure.

The base pressure of a foundation is the load strength transmitted from the bottom of the foundation

4.3 基底接触压力

基本上，每个结构都由土和岩石支撑，基础的主要功能是将结构的荷载传递到地基上。因此，分析基础与支撑地基之间的相互作用对结构的安全具有重要意义。

基底压力是从基础底部传递到地基土的荷载，也可以定义为

to the subsoil soil, and can also be defined as the contact pressure between the foundation and the soil under the foundation. The magnitude and distribution pattern of contact pressure depend on many factors such as the magnitude and distribution of structural loads, the stiffness and burial depth of foundations, and soil properties, such as soil type, density, and soil structure.

地基和基础之间的接触压力。接触压力的大小和分布模式取决于许多因素，如结构荷载的大小和分布、基础的刚度和埋深以及土体性质（土体类型、密度和土的结构）。

Figure 4.4 shows the measured base pressure distribution of a circular rigid foundation placed on sandy soil and hard cohesive soil.

图 4.4 显示了放置在砂土和硬黏性土上的圆形刚性基础的实测基底应力分布。

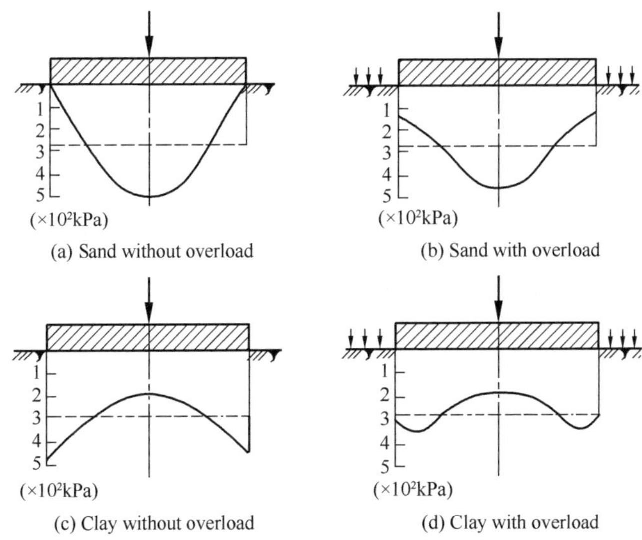

Figure 4.4 Distribution of contact pressure under foundation base

In Figure 4.4 (a), the foundation is placed on the surface of sandy soil with no overloading around, and the base pressure is distributed in a parabolic shape. This is because the sand particles at the edge of the foundation are easily squeezed out laterally, transferring the pressure they should bear to the middle part of the base. In Figure 4.4 (b), the foundation is also placed on the surface of sand, but there is a significant overload acting around it (equivalent to the foundation being buried deep), making it difficult to squeeze out the sand particles at the edge of the foundation. Therefore, the difference in reaction force between the central component of the foundation and the edge is much smaller than the former. If the rigid foundation model

在图 4.4(a) 中，基础放置在砂土表面，周围没有超载，基底压力呈抛物线分布。这是因为基础边缘的砂粒很容易横向挤出，从而将这些砂料应该承受的压力传递到基础的中间部分。在图 4.4(b) 中，基础也放置在砂表面上，但其周围有明显的超载作用（相当于基础埋得很深），使得基础边缘的砂粒难以挤出。因此，基础中心和边缘之间的反作用力差比前者小得多。如果刚性基础模型放置在硬黏土上，则测得的基础反力分布与放置在砂土

is placed on hard clay, the distribution map of the base reaction force measured is opposite to that when placed on sandy soil, showing a saddle shape with a small center and large edges. Due to the high cohesive force of hard clay, it is less likely to experience lateral extrusion of soil particles. Therefore, the difference in the distribution of base reaction forces between the cases of no overloading (Figure 4.4c) and overloading (Figure 4.4d) around the foundation is not as significant as that of sandy soil.

上时的分布相反，表现为中心小、边缘大的马鞍形。由于硬黏土具有高黏聚力，它不太可能受到土颗粒的侧向挤压。因此，基础周围无超载（图 4.4c）和超载（图 4.4d）情况下基础反力分布的差异不如砂土显著。

From the above discussion, it can be seen that the distribution of contact pressure is very complex. Therefore, in order to calculate the contact pressure, it is necessary to introduce assumptions.

从上面的讨论可以看出，接触压力的分布非常复杂。因此，为计算接触压力，有必要引入假设。

As we learned from the Saint Venant principle in elasticity, if a small portion of the force acting on the surface of an elastic body is replaced by another static equivalent system acting on the same part of the surface of the elastic body, the redistribution of this load produces substantial changes in local stress, but the effect on the stress at larger distances is negligible compared to the linear size of the surface on which the force changes. According to the Saint Venant principle, the calculation of soil stress based on the type of contact pressure distribution is limited to a given depth. When the depth range is exceeded, the correlation between the distribution of additional soil stress and the distribution pattern of contact pressure is small, while the contact pressure mainly depends on the direction and magnitude of the resultant force. Therefore, we can assume that the contact pressure under foundation base is uniformly distributed.

正如我们从弹性力学的圣维南原理中了解到的那样，如果作用在弹性体表面上的一小部分力被作用在该弹性体表面同一部分上的另一个静态等效系统所取代，则这种荷载的重新分布会导致局部应力发生实质性变化，但与力变化的表面的线性尺寸相比，在较大距离上对应力的影响可以忽略不计。根据圣维南原理，基于接触压力分布类型的土中应力计算仅限于给定的深度，当超过该深度范围时，附加应力的分布与接触压力的分布模式之间的相关性很小，而接触压力主要取决于合力的方向和大小。因此，可以假设基底的接触压力是均匀分布的。

4.3.1 Simplified Calculation of Contact Pressure

(1) Contact pressure under central load

The resultant force of the load on the foundation under the central load passes through the centroid of the foundation. Assuming a uniform distribution of

4.3.1 接触压力的简化计算

(1) 中心荷载作用下的接触压力

中心荷载作用下基础上的荷载合力通过基础的形心。假设基底压力均匀分布（图 4.5），平

base pressure (Figure 4.5), the average contact pressure p (kPa) (this load effect combination value is calculated according to the current national standard " Code for Design of Building Foundation" GB 50007—2011) can be calculated using the following equation:

均接触压力 p（kPa）（荷载效应组合值根据现行国家标准《建筑地基基础设计规范》GB 50007—2011 计算）可通过式（4.5）计算。

$$p=\frac{F+G}{A} \tag{4.5}$$

where F is vertical force acting on the foundation (kN); G is the total gravity of the foundation and its backfill soil (kN), $G=\gamma_G Ad$, where γ_G is the average unit weight of the foundation and backfill soil, generally taken as 20kN/m^3, but the unit weight below the groundwater level should be deducted by 10kN/m^3, and d is the burial depth of the foundation (m), which must be calculated from the design ground or the average design ground indoors and outdoors (Figure 4.5b); A is the area of foundation base (m^2), for rectangular base $A=lb$, l and b are the long side width (m) and short side width (m) of the rectangular base, respectively.

式中，F 为作用在基础上的竖向力（kN）；G 为基础及其上回填土的总重力（kN），$G=\gamma_G Ad$，其中 γ_G 为基础和回填土的平均重度，一般取 20kN/m^3，但地下水位以下的重度应扣除 10kN/m^3，d 为基础埋深（m），必须从设计地面或室内外平均设计地面计算（图 4.5b）；A 是基础底面积（m^2），对于矩形基底 $A=lb$，l 和 b 分别是矩形的长边宽度（m）和短边宽度（m）。

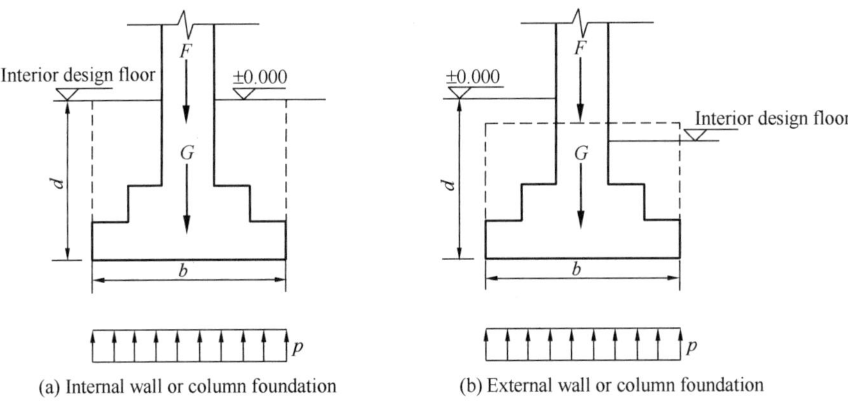

Figure 4.5 Distribution of contact pressure under foundation base

For a strip foundation with uniformly distributed loads, a unit length of strip is taken along the length direction to calculate the average contact pressure p. At this time, A in equation (4.5) is changed to b (m), and F and G are the corresponding values (kN/m) within the foundation strip.

(2) Contact pressure under eccentric load

对于荷载均匀分布的条形基础，沿长度方向取单位长度的条形来计算平均接触压力 p。此时，式（4.5）中的 A 变为 b（m），F 和 G 是条形基础内的相应值（kN/m）。

（2）偏心荷载作用下的接触压力

The rectangular foundation under unidirectional eccentric load is shown in Figure 4.6. When designing, the longitudinal direction of the base is usually aligned with the eccentric direction, and the maximum and minimum pressures p_{max} and p_{min} on both sides of the base (the combined value of this load effect is the same as under the action of central load) are calculated according to the equation for eccentric compression of short columns in material mechanics

单向偏心荷载作用下的矩形基础如图 4.6 所示。设计时，基底的纵向通常与偏心方向一致，基底两侧的最大 p_{max} 和最小压力 p_{min}（该荷载效应的组合值与中心荷载作用时相同）根据材料力学中短柱的偏心受压公式计算。

$$\left.\begin{matrix} p_{max} \\ p_{min} \end{matrix}\right\} = \frac{F+G}{lb} \pm \frac{M}{W} \tag{4.6}$$

where M is moment acting on the bottom surface of the rectangular foundation (kN·m); W is the resistance moment of the rectangular foundation base (m^3), $W = bl^2/6$. Substitute the eccentricity of the eccentric load $e = M/(F+G)$ (as shown in Figure 4.6) into Eq. 4.6 to obtain

式中，M 是作用在矩形基础底面上的力矩（kN·m）；W 为矩形基础底的抵抗矩（m^3），$W = bl^2/6$。将偏心荷载的偏心距 $e = M/(F+G)$（图 4.6）代入式（4.6），得式（4.7）。

$$\left.\begin{matrix} p_{max} \\ p_{min} \end{matrix}\right\} = \frac{F+G}{lb}\left(1 \pm \frac{6e}{l}\right) \tag{4.7}$$

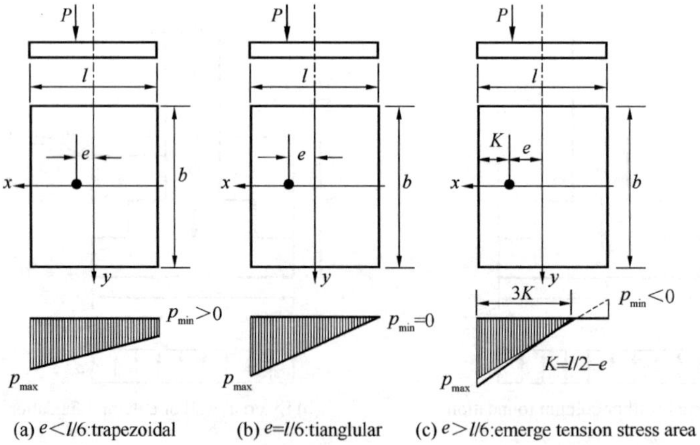

(a) $e<l/6$:trapezoidal (b) $e=l/6$:tianglular (c) $e>l/6$:emerge tension stress area

Figure 4.6 Distribution of rectangular contact pressure under unidirectional eccentric load

As can be seen from the above equation, when $e<l/6$, the distribution of contact pressure is trapezoidal (Figure 4.6a); When $e=l/6$, it forms a triangle (Figure 4.6b); When $e>l/6$, according to equation (4.7), the reaction force at the edge of the foundation far away from the eccentric load is negative, that is, $p_{min}<0$, as shown by the dashed line in Figure 4.6 (c). Due to the inability to withstand tensile forces between the

从上述公式可以看出，当 $e<l/6$时，接触压力的分布呈梯形（图 4.6a）；当 $e=l/6$ 时，接触压力分布变成一个三角形（图 4.6b）；当 $e>l/6$ 时，根据式（4.7），远离偏心荷载的基础边缘的反作用力为负，即 $p_{min}<0$，如图 4.6(c) 中的虚线所示。

base and foundation, the base and foundation partially detach, resulting in a redistribution of base pressure. Therefore, based on the condition that the eccentric load should be balanced with the base reaction force, the resultant force $F+G$ of the load should pass through the centroid of the triangular reaction force distribution diagram, as shown by the solid line in Figure 4.6 (c) . From this, the maximum pressure p_{max} at the edge of the base can be obtained

由于无法承受基底和基础之间的张力，基底和基础部分分离，导致基底压力重新分布。因此，在偏心荷载应与基础反力平衡的条件下，荷载的合力 $F+G$ 应通过三角形反力分布图的形心，如图 4.6 (c) 中的实线所示。由此，可得基底边缘的最大压力 p_{max}，即式 (4.8)。

$$p_{max}=\frac{2(F+G)}{3\left(\frac{l}{2}-e\right)b} \tag{4.8}$$

4.3.2 Additional Pressure at the Bottom of Foundation

Before the construction of a building, there is already self-weight stress in the soil, and the additional pressure at the bottom of foundation is the difference between the contact pressure and the self-weight stress in the soil before the construction of the foundation, which is the main cause of the additional stress and deformation of the foundation.

Generally, shallow foundations are always buried at a certain depth below the natural ground, where the vertical effective self-weight stress σ_{cd} in the original soil exists (Figure 4.7a). After excavation of the foundation pit, the original self-weight stress was eliminated and removed, indicating that the bottom of foundation was subjected to self-weight stress before construction (Figure 4.7b). The newly added average additional pressure at the bottom of foundation is obtained by subtracting the self-weight stress in the soil at the bottom of the foundation before construction from the average pressure of the building foundation after construction (Figure 4.7c) .

4.3.2 基底附加压力

建筑物施工前，土中已经存在自重应力，基础底部的附加压力是基底的接触压力和基础施工前土中的自重应力之差，这是引起地基附加应力和变形的主要原因。

一般来说，浅基础总是埋在天然地面以下一定深度，原状土中存在竖向有效自重应力 σ_{cd} (图 4.7a)。基坑开挖后，原始自重应力卸除，基底处在建前受自重应力作用 (图 4.7b)。建筑物建成后基底的平均压力减去建前基底处土的自重应力，才是新增加的基底平均附加压力 (图 4.7c)。

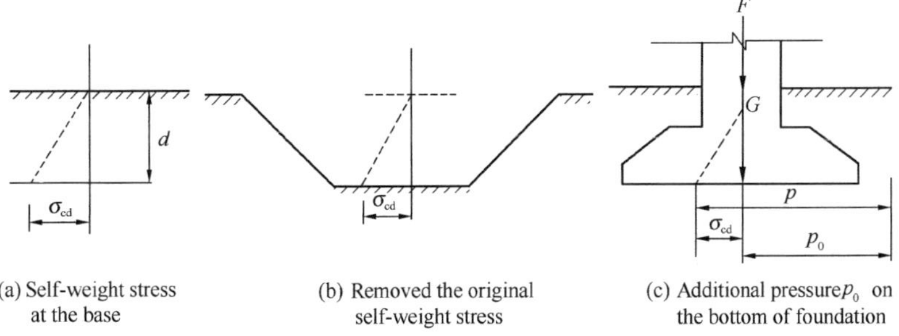

Figure 4.7 Analysis of additional pressure p_0 on the bottom of foundation

The average additional pressure p_0 on the bottom of foundation should be expressed according to the following equation:

基底的平均附加压力 p_0 应按式（4.9）表示。

$$p_0 = p - \sigma_{cd} = p - \gamma_m d \tag{4.9}$$

where p is the average contact pressure (kPa); γ_m is the weighted average unit weight of the natural soil layer above the base elevation (kN/m^3), in which the unit weight below the groundwater level should be taken as the floating unit weight; d is the foundation burial depth from natural ground (m).

式中，p 是平均接触压力(kPa)；γ_m 为基底标高以上天然土层的加权平均重度（kN/m^3)，其中地下水位以下应取浮重度；d 为距天然地面的基础埋深（m)。

4.4 Additional Stress in soil

4.4 土中附加应力

Additional stress is caused by external loads such as the weight of the building on the ground after its construction. "Additional" refers to increasing stress on the basis of self-weight stress. At present, in order to obtain the additional stress in soil generated by the additional load, the soil is generally assumed to be a continuous, uniform, and isotropic fully elastic material, while, it extends infinitely in both depth and horizontal directions, treating the subsoil as a homogeneous linear deformation half space, which can be directly solved according to the theory of elasticity.

附加应力是由外部荷载引起的，例如建筑物建成后在地面上的重量。“附加”是指在自重应力的基础上增加的压力。目前，为了获得附加荷载在土中产生的附加应力，通常假设土是连续、均匀、各向同性的完全弹性材料，且在深度和水平方向上无限延伸，将地基视为均匀的线性变形半空间，可以根据弹性理论直接求解。

4.4.1 Additional Stress in Soil Under Vertical Concentrated Force

4.4.1 竖向集中力作用下的附加应力

In 1885, Boussinesq proposed an elastic mechanics solution for the stress and displacement caused at any point in an elastic half space when a vertically concentrated force is applied on the surface of the half space. As shown in Figure 4.8, the solutions for the six stress components and three displacement components at any point M (x, y, z) in a half space (equivalent to a ground surface) are as follows:

1885 年，布辛奈斯克提出了一个弹性力学解，用于求解竖向集中力施加在半空间表面时弹性半空间中任意点处产生的应力和位移。如图 4.8 所示，半空间（相当于地面）中任意点 $M(x,y,z)$ 处的 6 个应力分量和 3 个位移分量的解见式（4.10）～式（4.18）。

$$\sigma_x = \frac{3P}{2\pi} \cdot \left\{\frac{x^2 z}{R^5} + \frac{1-2\mu}{3}\left[\frac{1}{R(R+z)} - \frac{(2R+z)x^2}{(R+z)^2 R^3} - \frac{z}{R^3}\right]\right\} \tag{4.10}$$

$$\sigma_y = \frac{3P}{2\pi} \cdot \left\{\frac{y^2 z}{R^5} + \frac{1-2\mu}{3}\left[\frac{1}{R(R+z)} - \frac{(2R+z)y^2}{(R+z)^2 R^3} - \frac{z}{R^3}\right]\right\} \tag{4.11}$$

$$\sigma_z = \frac{3P}{2\pi} \cdot \frac{z^3}{R^5} = \frac{3P}{2\pi R^2} \cos^3\beta \tag{4.12}$$

$$\tau_{xy} = \tau_{yx} = \frac{3P}{2\pi} \cdot \left[\frac{xyz}{R^5} + \frac{1-2\mu}{3} \cdot \frac{(2R+z)xy}{(R+z)^2 R^3}\right] \tag{4.13}$$

$$\tau_{zy} = \tau_{yz} = -\frac{3P}{2\pi} \cdot \frac{yz^2}{R^5} = -\frac{3Py}{2\pi R^3} \cos^2\beta \tag{4.14}$$

$$\tau_{zx} = \tau_{xz} = -\frac{3F}{2\pi} \cdot \frac{xz^2}{R^5} = -\frac{3Px}{2\pi R^3} \cos^2\beta \tag{4.15}$$

$$u = \frac{P(1+\mu)}{2\pi E}\left[\frac{xz}{R^3} - (1-2\mu)\frac{x}{R(R+z)}\right] \tag{4.16}$$

$$v = \frac{P(1+\mu)}{2\pi E}\left[\frac{yz}{R^3} - (1-2\mu)\frac{y}{R(R+z)}\right] \tag{4.17}$$

$$w = \frac{P(1+\mu)}{2\pi E}\left[\frac{z^2}{R^3} - 2(1-\mu)\frac{1}{R}\right] \tag{4.18}$$

where σ_x, σ_y, and σ_z are normal stresses parallel to the x, y, and z coordinate axes respectively; τ_{xy}, τ_{yz}, τ_{zx} are shear stress, where the first subscript represents the coordinate axis parallel to the normal direction of the micro surface it acts on, and the second subscript represents the coordinate axis parallel to the direction of its action; u, v and w are the displacement of point M along the x, y, and z directions of the coordinate axis; P is vertical concentrated force acting on the coordinate origin O; R is the distance from point M to the coordinate origin O, $R=(x^2+y^2+z^2)^{0.5}=(r^2+z^2)^{0.5}=z/\cos\beta$; β is the angle between the line R and the z-axis; r is the horizontal distance between point M and the point of concentrated force application; E is elastic modulus (or deformation modulus of soil specifically used in soil mechanics, represented by E_0); μ is Poisson's ratio.

式中，σ_x、σ_y 和 σ_z 分别是平行于 x、y 和 z 坐标轴的法向应力；τ_{xy}，τ_{yz}，τ_{zx} 为剪应力，其中第一个下标表示平行于其作用的微面法线方向的坐标轴，第二个下标表示垂直于其作用方向的坐标轴线；u、v 和 w 是点 M 沿坐标轴的 x、y 和 z 方向的位移；P 是作用在坐标原点 O 上的竖向集中力；R 是从点 M 到坐标原点 O 的距离，$R=(x^2+y^2+z^2)^{0.5}=(r^2+z^2)^{0.5}=z/\cos\beta$；$\beta$ 是线 R 和 z 轴之间的角度；r 是点 M 和集中力施加点之间的水平距离；E 为弹性模量（或土力学中专门使用的土的变形模量，用 E_0 表示）；μ 是泊松比。

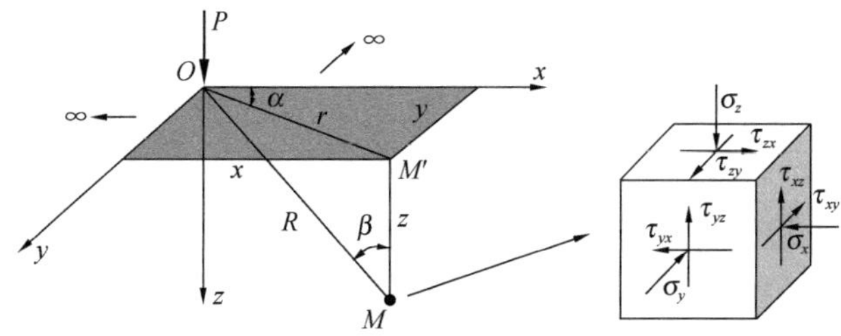

Figure 4.8 Stress in soil caused by vertical concentrated force

If $R=0$ is substituted into the above equations, the results obtained are all infinite. Therefore, the selected calculation point should not be too close to the point of action of the concentrated force.

如果将 $R=0$ 代入上述方程，所得结果均为无穷大。因此，所选的计算点不应太靠近集中力的作用点。

If the distance between a point M in the subsoil and the local load is much larger than the size of the load surface, a concentrated force P can be used instead of the local load, and then Eq. (4.12) can be directly applied to calculate the σ_z of that point. For the convenience of calculation, $R=(r^2+z^2)^{0.5}$ can be substituted into Eq. (4.12), then

如果地基中的点 M 与局部荷载之间的距离远大于荷载面的尺寸，则可以使用集中力 P 代替局部荷载，可直接用式（4.12）来计算该点的 σ_z。为便于计算，可以将 $R=(r^2+z^2)^{0.5}$ 代入式（4.12），得式（4.19）。用式（4.20）表示 α，

$$\sigma_z=\frac{3}{2\pi}\cdot\frac{1}{[(r/z)^2+1]^{5/2}}\cdot\frac{P}{z^2} \tag{4.19}$$

Let

$$\alpha=\frac{3}{2\pi}\cdot\frac{1}{[(r/z)^2+1]^{5/2}} \tag{4.20}$$

Then the above equation can be rewritten as

式（4.19）可改写为式（4.21）。

$$\sigma_z=\alpha(P/z^2) \tag{4.21}$$

where α is the vertical additional stress coefficient of the subsoil under the action of concentrated force P, abbreviated as concentration stress factor.

式中，α 是集中力 P 作用下竖向附加应力系数，缩写为集中应力系数。

Several vertical concentrated forces P_i ($i=1, 2, \dots, n$) act on the surface of the subsoil. According to the equivalent load replacement method, which is based on the superposition principle, the additional stress σ_z at a certain point M at a depth of z below the ground should be the sum of the additional stresses caused by each concentrated force acting alone at point M, that is

多个竖向集中力 P_i（$i=1, 2, \cdots, n$）作用于地基表面。根据基于叠加原理的等效荷载置换法，地基 z 深度处某点 M 的附加应力 σ_z 应为单独作用在 M 点的每个集中力引起的附加应力之和，即式（4.22）所示。

$$\sigma_z=\sum_{i=1}^{n}\alpha_i\frac{P_i}{z^2}=\frac{1}{z^2}\sum_{i=1}^{n}\alpha_i P_i \tag{4.22}$$

where α_i is the i-th concentrated stress coefficient, and in the calculation, r_i is the horizontal distance from the i-th concentrated load point to point M.

式中，α_i 是第 i 个集中应力系数，在计算中，r_i 是从第 i 个集中力作用点到点 M 的水平距离。

When the planar shape or distribution of local loads is irregular, the load surface (or foundation base) can be divided into several regularly shaped (such as rectangular) unit areas, and the distributed load on each unit area can be approximated by the concentrated

当平面形状或局部荷载分布不规则时，荷载表面（或基础底面）可以划分为几个规则形状（如矩形）的单元，每个单元上的分布荷载可以通过作用在单元

force acting on the centroid of the unit area. In this way, Eq. (4.22) can be used to calculate the additional stress at a certain point M in the subsoil. Due to the infinite value of σ_z near the point of concentrated force, this method is not suitable for calculation points that are too close to the load surface. Its calculation accuracy depends on the size of the unit area. Generally, when the long side of the rectangular unit area is less than 1/2, 1/3, or 1/4 of the distance from the centroid of the area to the calculation point, the error in calculating the additional stress is not greater than 6%, 3%, or 2%, respectively.

区域形心上的集中力来近似表示。这样，就可以用式（4.22）计算地基中某一点 M 处的附加应力。由于集中力点附近 σ_z 的无穷大，该方法不适用于过于靠近荷载面的计算点。其计算精度取决于单元面积的大小。通常，当矩形单元面积的长边小于面积形心到计算点距离的 1/2、1/3 或 1/4 时，计算附加应力的误差分别不大于 6%、3%或 2%。

Example 4.2

A certain strip foundation, as shown in Figure 4.9. Apply a concentrated load of $F=340\text{kN/m}$ on the foundation, calculate the additional stress of the foundation below the midpoint, and draw a distribution diagram of the additional stress.

例题 4.2

如图 4.9 所示的某条形基础。在基础上施加集中荷载 $F=340\text{kN/m}$，计算中心点以下地基的附加应力，并绘制附加应力分布图。

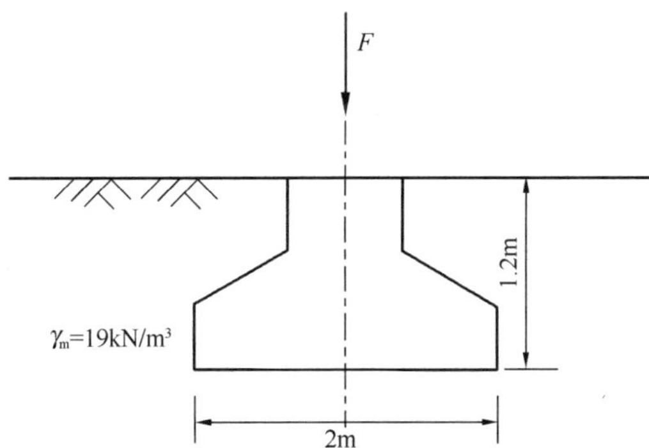

Figure 4.9 Strip foundation

Solution

解答

$$p=\frac{F+G}{A}=\frac{340+1\times 2\times 1.2\times 20}{1\times 2}=194\text{kPa}$$

$$p_0=p-\gamma_m d=194-19\times 1.2=171.2\text{kPa}$$

Substitute the value of p_0 into Eq. (4.19) and calculate the vertical additional stress in the subsoil for $z=0$m, 1m, 2m, 4m, 6m, and 8m, respectively. The results are shown in Figure 4.10.

将 p_0 的值代入式（4.19），分别计算 $z=0$m、1m、2m、4m、6m 和 8m 时地基中的竖向附加应力，结果如图 4.10 所示。

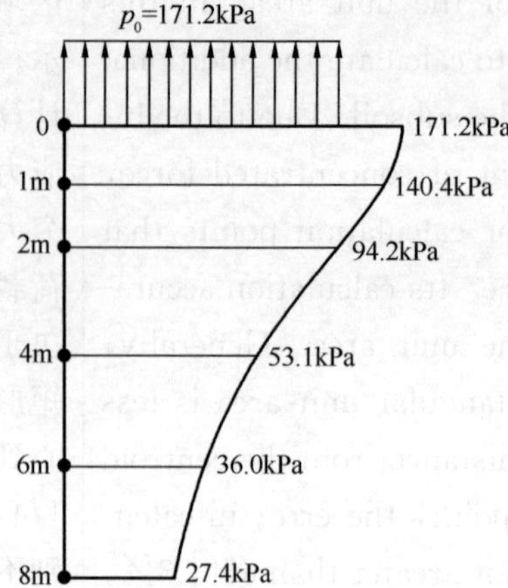

Figure 4.10　Distribution of vertical stress along the depth

4.4.2　Additional Stress in Soil Under Distributed Load

The load acting on the foundation of a building is always a local load distributed over a certain area, so theoretically there is no concentrated force in practice. However, according to the superposition principle of elastic mechanics and using the Boussinesq solution, the additional stress in the subsoil with arbitrary distribution and irregular load surface shape can be obtained through the equivalent load method, or the additional stress in the subsoil under various local loads can be directly solved by integration.

As shown in Figure 4.11, a semi-infinite elastic space surface is subjected to a load $p(x, y)$ with an arbitrary shape distribution area. Establish a coordinate system as shown in Figure 4.11.

Take a small load distribution area $dA = d\varepsilon d\eta$, the force on this small area can be expressed as:

4.4.2　分布荷载作用下的土中附加应力

作用在建筑物基础上的荷载总是分布在一定区域上的局部荷载，因此理论上，实际中没有集中力。然而，根据弹性力学的叠加原理，并使用布辛奈斯克方程，可以通过等效荷载法获得任意分布和不规则荷载表面形状下地基中的附加应力，或者可以通过积分直接求解各种局部荷载下地基中的附加应力。

如图 4.11 所示，半无限弹性空间表面受到任意形状分布区域的荷载 $p(x, y)$。建立如图 4.11所示的坐标系。

取一个小的荷载分布区域 $dA = d\varepsilon d\eta$，该小区域上的力可表示为式（4.23）。

$$dP = p(x, y) d\varepsilon d\eta \tag{4.23}$$

Due to its small magnitude, it can be considered as a concentrated force. Therefore, substituting it into Boussinesq Eq. (4.12) yields.

由于其幅值较小，可以将其视为集中力。因此，将其代入 Boussinesq 方程［式（4.12）］可得式（4.24）。

$$d\sigma_z = \frac{3z^3}{2\pi} \frac{dP}{[(x-\varepsilon)^2 + (y-\eta)^2 + z^2]^{5/2}} \tag{4.24}$$

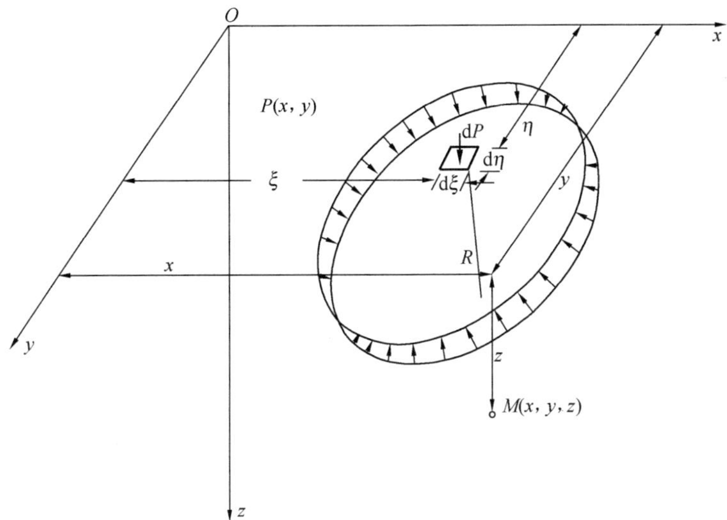

Figure 4.11 Vertical additional stress analysis in subsoil under distributed load

Integrating the above equation over the entire load distribution area yields

将式（4.24）在整个荷载分布面积上积分可得式（4.25）。

$$\sigma_z = \iint_A \mathrm{d}\sigma_z = \frac{3z^3}{2\pi}\iint_A \frac{p(x,y)\mathrm{d}x\mathrm{d}y}{[(x-\xi)^2+(y-\eta)^2+z^2]^{\frac{5}{2}}} \tag{4.25}$$

The above equation is the expression for the vertical additional stress in the soil induced by any distributed load on any shape distribution area.

该式为任意形状分布面积的荷载引起的竖向附加应力的表达式。

In Eq. (4.25), if the integrable function is integrable, the functional expression for the vertical additional stress can be obtained. On the contrary, we can also use numerical integration to obtain the results.

在式（4.25）中，如果被积函数是可积的，则可以获得竖向附加应力的函数表达式。如果不可积，也可使用数值积分来获得结果。

Due to the assumption that the contact pressure under foundation base is linearly distributed, the integrable function is integrable for regular shaped foundations such as rectangular foundations, circular foundations, and strip foundations. By analyzing the integrand function, it can be concluded that the integration result can definitely be expressed as

由于假设基底接触压力呈线性分布，因此对于矩形基础、圆形基础和条形基础等规则形状的基础，被积函数是可积的。通过分析被积函数，可以得出结论，积分结果可以确定地表示为式（4.26）。

$$\sigma_z = \alpha \cdot p_0 \tag{4.26}$$

where α is the vertical additional stress coefficient. It is only related to the coordinates of the observation point and dimensions of foundation, that is, it is a function of

式中，α 是竖向附加应力系数。它只与观测点的坐标和基础的尺寸有关，即它是观测点位置和基

the position of the observation point and dimensions of foundation.

础尺寸的函数。

(1) Uniformly distributed rectangular load

As shown in Figure 4.12, assuming the long and short widths of the rectangular load surface are l and b, respectively, the vertically uniformly distributed load p (or the average additional pressure p_0 on the foundation bottom) acting on the surface of the elastic half space.

（1）均布矩形荷载

如图 4.12 所示，假设矩形荷载面的长边和短边分别为 l 和 b，竖向均匀分布荷载 p（或基底的平均附加压力 p_0）作用在弹性半空间表面上。

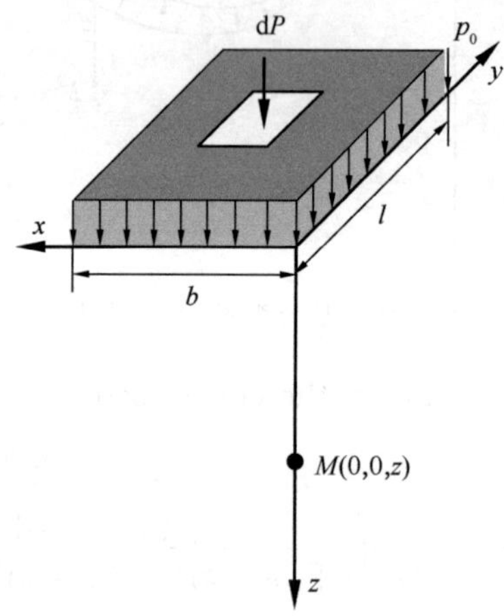

Figure 4.12 Vertical additional stress analysis in subsoil under distributed rectangular load

First, use the integration method to obtain the additional stress of the subsoil at any depth z under the corner of the rectangular load surface, and then use the corner method to obtain the additional stress of the subsoil at any point under the rectangular load. Taking the corner point of the rectangular load surface as the coordinate origin, taking a micro element area dP within the load surface, and replacing the uniformly distributed load on it with a concentrated force, the vertical additional stress $\mathrm{d}\sigma_z$ caused by the concentrated force at the M point at any depth z below the corner point o can be obtained according to Eq. (4.24) that

首先，使用积分法获得矩形荷载表面角点下任意深度 z 处的地基附加应力，然后用角点法获得矩形荷载下任意点处的地基附加应力。以矩形荷载面的角点为坐标原点，取荷载面内的微面积 dP，用集中力代替其上均匀分布的荷载，根据式（4.24），可以获得角点 o 下方任意深度 z 处 M 点集中力引起的竖向附加应力 $\mathrm{d}\sigma_z$，即式（4.27）所示。

$$\sigma_z=\iint_A \mathrm{d}\sigma_z=\iint_A \frac{3\mathrm{d}P}{2\pi}\frac{z^3}{R^5}=\frac{3p_0z^3}{2\pi}\int_0^l\int_0^b\frac{1}{(x^2+y^2+z^2)^{5/2}}\mathrm{d}x\mathrm{d}y=\alpha_c p_0 \tag{4.27}$$

where α_c is the vertical additional stress coefficient under uniformly distributed rectangular load corner points, abbreviated as corner point stress factor. It is a function of l/b and z/b. It can be obtained by checking

式中，α_c是均匀分布矩形荷载角点下的竖向附加应力系数，缩写为角点应力系数。它是 l/b 和 z/b 的函数。根据 l/b 和 z/b 的值，

the additional stress coefficient table in the " Code for Design of Building Foundation" GB 50007—2011 based on the values of l/b and z/b.

可以通过《建筑地基基础设计规范》GB 50007—2011 中的附加应力系数表来获得 α_c。

For the case where the calculation point of additional stress for uniformly distributed rectangular loads is not at the corner point, Eq. (4.27) can be used to obtain it using the corner points method. Four cases are listed in Figure 4.13 where the calculation point is not located at the corner of the rectangular load surface (at any depth z below point o in the figure). When calculating, the load surface is divided into several rectangular areas by point o, so that point o must be the common corner point of each divided rectangle. Then, according to Eq. (4.27), the additional stress σ_z at the same depth z at each corner point of the rectangle is calculated, and its algebraic sum is obtained.

对于均匀分布矩形荷载的附加应力计算点不在角点的情况，可以使用式（4.27）通过角点法获得。图 4.13 列出了四种情况，其中计算点不位于矩形荷载面的角点处（在图中点 o 下方的任意深度 z 处）。计算时，荷载面由点 o 划分为几个矩形区域，因此点 o 必须是每个划分矩形的公共角点。然后，根据式（4.27），计算矩形每个角点相同深度 z 处的附加应力 σ_z，并得到其代数和。

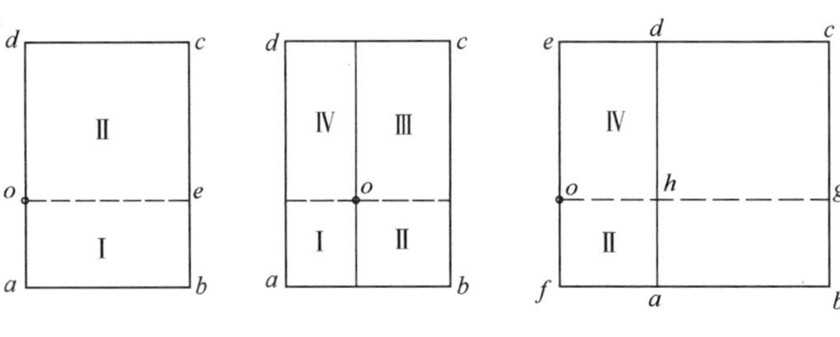

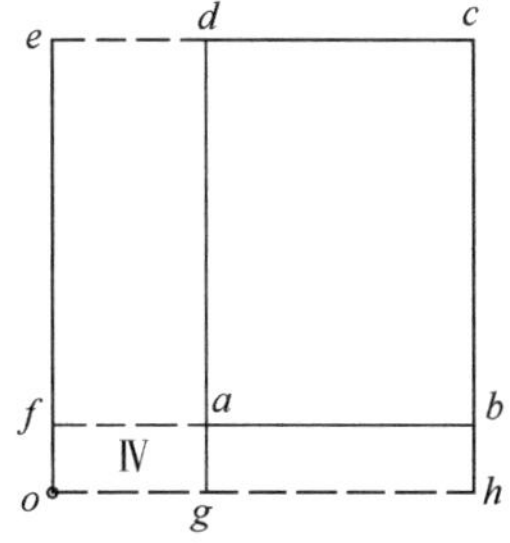

Figure 4.13 Calculation of additional stress on foundation under uniformly distributed rectangular load using corner method

The formulas for the four situations are as follows:

Case 1: Point o is located at the edge of the load surface (Figure 4.13a)

以下是四种非角点下的附加应力计算：

情况 1：点 o 位于荷载面的边缘（图 4.13a）

$$\sigma_z = (\alpha_{c\,\mathrm{I}} + \alpha_{c\,\mathrm{II}}) p_0$$

where $\alpha_{c\,\mathrm{I}}$ and $\alpha_{c\,\mathrm{II}}$ represent the corner stress coefficients corresponding to the areas Ⅰ and Ⅱ, respectively.

式中，$\alpha_{c\,\mathrm{I}}$ 和 $\alpha_{c\,\mathrm{II}}$ 分别表示与区域Ⅰ和Ⅱ对应的角应力系数。

Case 2: Point o is located within the load plane (Figure 4.13b)

情况 2：点 o 位于荷载面内（图 4.13b）

$$\sigma_z = (\alpha_{c\text{I}} + \alpha_{c\text{II}} + \alpha_{c\text{III}} + \alpha_{c\text{IV}}) p_0$$

If point o is located at the center of the load surface, then $\alpha_{c\text{I}} = \alpha_{c\text{II}} = \alpha_{c\text{III}} = \alpha_{c\text{IV}}$, obtaining $\sigma_z = 4\alpha_{c\text{I}} p_0$, that is, using the corner method to find the solution of σ_z at the center point of a uniformly distributed rectangular load surface.

如果点 o 位于荷载面的中心，则 $\alpha_{c\text{I}} = \alpha_{c\text{II}} = \alpha_{c\text{III}} = \alpha_{c\text{IV}}$，得到 $\sigma_z = 4\alpha_{c\text{I}} p_0$，即使用角点法求解均匀分布矩形荷载面的中心点处的 σ_z。

Case 3: Point o is located outside the edge of the load surface (Figure 4.13c)

情况 3：点 o 位于荷载面的边缘外（图 4.13c）

At this point, the load surface *abcd* can be considered as a combination of the difference between Ⅰ (*ofbg*) and Ⅱ (*ofah*) and the difference between Ⅲ (*oecg*) and Ⅳ (*oedh*). Therefore,

此时，荷载面 *abcd* 可以被视为Ⅰ(*ofbg*)和Ⅱ(*ofah*)之间的差值以及Ⅲ(*oecg*)和Ⅳ(*oedh*)之间的差值的和。因此，

$$\sigma_z = (\alpha_{c\text{I}} - \alpha_{c\text{II}} + \alpha_{c\text{III}} - \alpha_{c\text{IV}}) p_0$$

Case 4: Point o is located outside the corner of the load surface (Figure 4.13d)

情况 4：点 o 位于荷载面的角点外（图 4.13d）

The load surface is considered to be formed by subtracting Ⅱ (*ohbf*) and Ⅲ (*ogde*) from the areas of Ⅰ (*ohce*) and Ⅳ (*ogaf*). Therefore,

荷载面是从Ⅰ(*ohce*)和Ⅳ(*ogaf*)的面积中减去Ⅱ(*ohbf*)和Ⅲ(*ogde*)，因此，

$$\sigma_z = (\alpha_{c\text{I}} - \alpha_{c\text{II}} - \alpha_{c\text{III}} + \alpha_{c\text{IV}}) p_0$$

The vertical additional stress σ_z value at any point in the uniformly distributed strip load surface foundation can also be calculated using the corner method using Eq. (4.27). The corner stress coefficient α_c in the equation is taken as $l/b=10$, with an error not exceeding 0.005.

线性均布荷载作用下基础表面任一点的竖向附加应力 σ_z 值也可以使用式（4.27）的角点法计算。式中的角点应力系数 α_c 取 $l/b=10$，误差不超过 0.005。

(2) Uniformly distributed triangular load

（2）均布的三角形荷载

If the vertical load acting on the surface of an elastic half space is distributed in a triangular shape along one side b of the rectangular area (the load distribution along the other side l remains unchanged), and the maximum value of the load is p_0, and the corner point 1 of the zero value side of the load is taken as the coordinate origin (Figure 4.14), then the distributed load on the micro element area $\mathrm{d}x\mathrm{d}y$ at a certain point (x, y) in the load plane can be replaced by the concentrated force $x/bp_0 \mathrm{d}x\mathrm{d}y$.

如果作用在弹性半空间表面上的竖向荷载沿矩形区域的短边 b 呈三角形分布（沿另一侧 l 的荷载分布保持不变），荷载的最大值为 p_0，以荷载零值侧的角点 1 为坐标原点（图 4.14），则荷载平面中某一点 (x, y) 处微元区域 $\mathrm{d}x\mathrm{d}y$ 上的分布荷载可以用集中力 $x/bp_0 \mathrm{d}x\mathrm{d}y$ 代替。

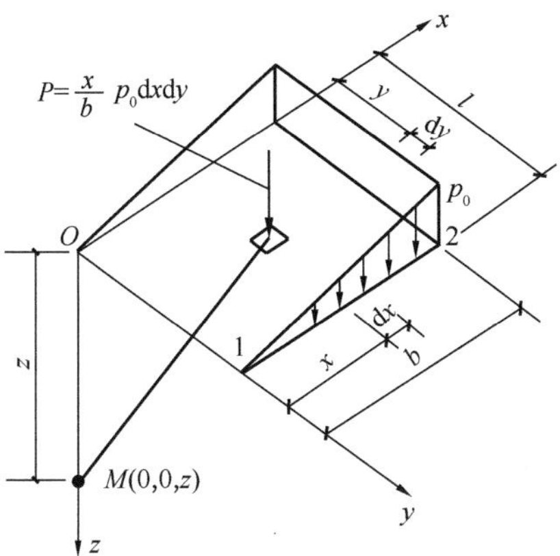

Figure 4.14 Additional stress at the corner points of a rectangular load surface with a triangular distribution

The additional stress σ_z caused by the concentrated force at point M at depth z below corner point 1 is expressed by Eq. (4.12)

用式（4.12）表示角点 1 下方深度 z 处的点 M 处的集中力引起的附加应力 σ_z 可得式（4.28）。

$$\sigma_z = \iint_A \mathrm{d}\sigma_z = \iint_A \frac{3}{2\pi} \frac{p_0 x z^3}{bR^5} \mathrm{d}x\mathrm{d}y = \frac{3p_0 z^3}{2\pi b} \int_0^l \int_0^b \frac{1}{(x^2+y^2+z^2)^{5/2}} \mathrm{d}x\mathrm{d}y = \alpha_{t1} p_0 \quad (4.28)$$

Similarly, the vertical additional stress σ_z at any depth z below the corner point 2 of the maximum load edge can also be calculated as

同样，最大荷载边缘角点 2 下方任意深度 z 处的竖向附加应力 σ_z 可表达为式（4.29）。

$$\sigma_z = \alpha_{t2} \cdot p_0 = (\alpha_c - \alpha_{t1}) p_0 \quad (4.29)$$

where α_{t1} and α_{t2} are respectively additional stress coefficient under corner 1 and corner 2 of a rectangular load surface with triangular distribution. They are functions of l/b and z/b. It can be obtained by checking the additional stress coefficient table in the " Code for Design of Building Foundation" GB 50007—2011 based on the values of l/b and z/b.

式中，α_{t1} 和 α_{t2} 分别是三角形分布的矩形荷载面的角点 1 和角点 2 处的附加应力系数，它们是 l/b 和 z/b 的函数，根据 l/b 和 z/b 的值，可通过《建筑地基基础设计规范》GB 50007—2011 中的附加应力系数表来获得。

Using the additional stress coefficients α_c, α_{t1} and α_{t2} at the corners of rectangular load surfaces with uniform and triangular distributions, the vertical additional stress σ_z at any point in the soil with trapezoidal distribution can be calculated using the corner method; The vertical additional stress in the soil can also be calculated for uniformly distributed, triangular, or trapezoidal strip load surfaces.

运用上述均布和三角形分布的矩形荷载面角点下的附加应力系数 α_c、α_{t1} 和 α_{t2}，即可用角点法求梯形分布时土中任意点的竖向附加应力 σ_z；亦可计算均布、三角形或梯形分布的面荷载作用下（l/b=10）的土中竖向附加应力。

(3) Uniform circular load

Assuming the radius of the circular load area is r_0 and the vertically uniformly distributed load acting on the surface of the elastic half space is p, taking the center point of the circular load surface as the coordinate origin o (Figure 4.15), and taking the micro area $dA = rd\theta dr$ on the load area, replacing the uniformly distributed circular load on the micro area in Figure 4.15 with the concentrated force $p_0 dA$, Eq. (4.12) can be used to calculate the σ_z at any depth z below the midpoint of the uniformly distributed circular load surface using the integration method.

(3) 均布的圆形荷载

假设圆形荷载区域的半径为 r_0，作用在弹性半空间表面上的垂直均匀分布的荷载为 p，以圆形荷载面的中心点为坐标原点 O（图 4.15），取荷载区域上的微区域 $dA = rd\theta dr$，用集中力 $p_0 dA$ 代替图 4.15 中微区域上的均布的圆形荷载，则可用式（4.12）积分计算均布圆形荷载面中点以下任意深度 z 处的 σ_z。

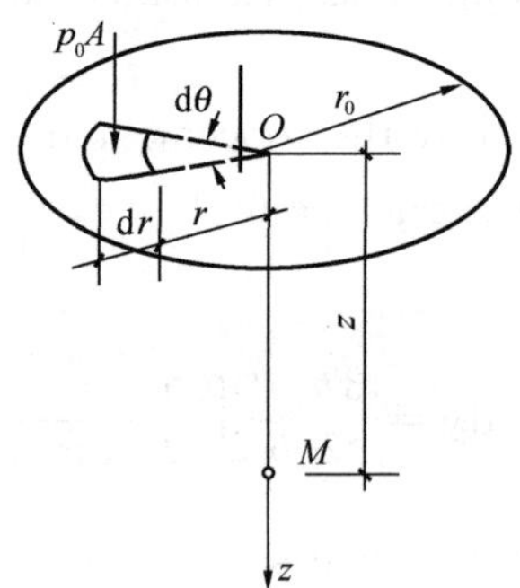

Figure 4.15 Vertical additional stress below the midpoint of a uniformly distributed circular load surface

The σ_z at point M at any depth z below the midpoint of the load surface can be obtained as follows:

荷载面中点以下任意深度 z 处 M 点的 σ_z 可按式（4.30）计算。

$$\sigma_z = \iint_A d\sigma_z = \frac{3p_0 z^3}{2\pi}\int_0^{2\pi}\int_0^{r_0}\frac{rd\theta dr}{(r^2+z^2)^{5/2}} = \alpha_r p_0 \tag{4.30}$$

where α_r is the additional stress coefficient at the center point of a uniformly distributed circular load surface, which is a function of z/r_0. It can be obtained by checking the additional stress coefficient table in the " Code for Design of Building Foundation" GB 50007—2011 based on the values of z/r_0.

式中，α_r 是均布的圆形荷载面中心点处的附加应力系数，它是 z/r_0 的函数。根据 z/r_0 的值查《建筑地基基础设计规范》GB 50007—2011 中的附加应力系数表可得。

4.4.3 Additional Stress in Soil Under the Line Load and Strip Load

A strip load of infinite length is applied on the surface of an elastic half space, and the load can be distributed in any form along the width, but remains

4.4.3 线荷载和条形荷载下的土中附加应力

在弹性半空间的表面上施加无限长的条形荷载，荷载可以沿宽度以任何形式分布，但沿长度

unchanged along the length direction. At this point, the stress state generated in the soil belongs to a plane strain problem. Therefore, for strip foundations such as wall foundations, retaining wall foundations, roadbeds, dam foundations, etc. , in order to solve the additional stress of the soil under strip loads, the following will first introduce the solution under linear loads.

方向保持不变。此时，土中产生的应力状态属于平面应变问题。为了求解条形基础（如墙基础、挡土墙基础、路基、坝基等）在条形荷载下土中的附加应力，下面先介绍线荷载作用下的解。

(1) Line load

Line load is a uniformly distributed load on an infinitely long straight line on the surface of an elastic half space. As shown in Figure 4. 16, a vertical line load $\bar{p}$ (kN/m) acts on the y-axis, and the distributed load on a certain differential segment dy along the y-axis is replaced by the concentrated force $P=\bar{p}\mathrm{d}y$.

(1) 线荷载

线荷载是弹性半空间表面上无限长直线上均匀分布的荷载。如图 4. 16 所示，竖向线荷载 $\bar{p}$（kN/m）作用在 y 轴上，沿 y 轴在某个微段 dy 上的分布荷载被集中力 $P=\bar{p}\mathrm{d}y$ 所取代。

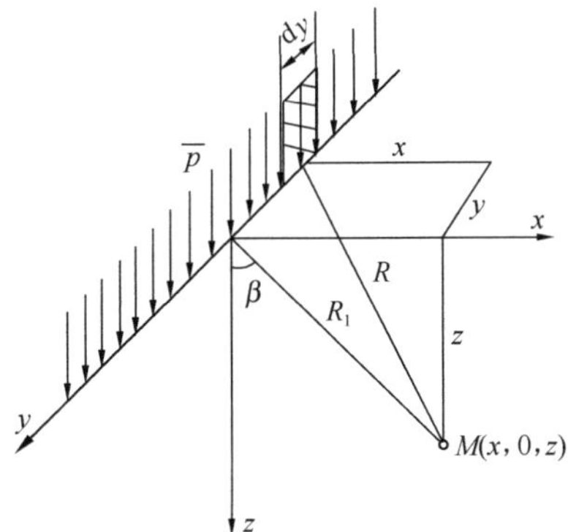

Figure 4. 16 Additional stress in subsoil under line load

By substitute $p=\bar{p}\mathrm{d}y$ into Eq. (4. 12) and integrating over the entire length of the negative and positive infinite intervals, the expression of vertical additional stress at any point M in the soil under line load can be obtained.

将 $P=\bar{p}\mathrm{d}y$ 代入式（4. 12）并在整个长度方向上从负无穷到正无穷进行积分，可得线荷载下土中任意点 M 处的竖向附加应力表达式（4. 31）。

$$\sigma_z=\int_{-\infty}^{+\infty}\mathrm{d}\sigma_z=\int_{-\infty}^{+\infty}\frac{3\bar{p}z^3}{2\pi R^5}\mathrm{d}y=\frac{2\bar{p}}{\pi R_1}\cos^3\beta \tag{4.31}$$

Similarly, it can be derived that

同样，可得水平方向附加应力式（4. 32）和剪切附加应力式（4. 33）。

$$\sigma_x=\int_{-\infty}^{+\infty}\mathrm{d}\sigma_x=\frac{2\bar{p}}{\pi R_1}\cos\beta\sin^2\beta \tag{4.32}$$

$$\tau_{xz}=\tau_{zx}=\frac{2\bar{p}}{\pi R_1}\cos^2\beta\sin\beta \tag{4.33}$$

Due to the uniform distribution and infinite extension of the line load along the y-axis, the stress state on any plane perpendicular to the y-axis is completely the same, which belongs to the plane strain problem. Therefore,

由于线荷载沿 y 轴均匀分布并无限延伸，垂直于 y 轴的任何平面上的应力状态都是完全相同的，属于平面应变问题。因此，

$$\tau_{xy} = \tau_{yx} = \tau_{yz} = \tau_{zy} = 0$$

$$\sigma_y = \mu(\sigma_x + \sigma_z)$$

(2) Strip load

If a vertical strip load is uniformly distributed along the width direction (Figure 4.17), then the uniformly distributed strip load p_0 (kN/m^2) can be replaced by a line load $\bar{p}$ on a certain differential segment dx along the x-axis.

（2）条形荷载

如果竖向条形荷载沿宽度方向均匀分布（图 4.17），则均布的条形荷载 p_0（kN/m^2）可以用沿 x 轴的某个微段 dx 上的线荷载 $\bar{p}$ 代替。

$$\bar{p} = p_0 \mathrm{d}x = \frac{p_0 R_1}{\cos\beta} \mathrm{d}\beta$$

By substituting it into Eq. (4.31) and integrating it over the entire width, we can obtain

将替代的线荷载代入式（4.31）并在整个宽度上积分，可得式（4.34）。

$$\sigma_z = \frac{p_0}{\pi}[\sin\beta_2\cos\beta_2 - \sin\beta_1\cos\beta_1 + (\beta_2 - \beta_1)] = \alpha_{sz} p_0 \tag{4.34}$$

Similarly, it can be obtained that

同样，可得式（4.35）和式（4.36）。

$$\sigma_x = \frac{p_0}{\pi}[-\sin(\beta_2 - \beta_1)\cos(\beta_2 + \beta_1) + (\beta_2 - \beta_1)] = \alpha_{sx} p_0 \tag{4.35}$$

$$\tau_{xz} = \tau_{zx} = \frac{p_0}{\pi}(\sin^2\beta_2 - \sin^2\beta_1) = \alpha_{sxz} p_0 \tag{4.36}$$

where α_{sz}, α_{sx} and α_{sxz} are the corresponding three additional stress coefficients under uniformly distributed strip loads, all of which are functions of z/b and x/b. It can be obtained by checking the additional stress coefficient table in the " Code for Design of Building Foundation" GB 50007—2011 based on the values of z/b and x/b.

式中，α_{sz}、α_{sx} 和 α_{sxz} 是均布条形荷载下相应的三个附加应力系数，均为 z/b 和 x/b 的函数。根据 z/b 和 x/b 的值，查《建筑地基基础设计规范》GB 50007—2011 中的附加应力系数值表可得。

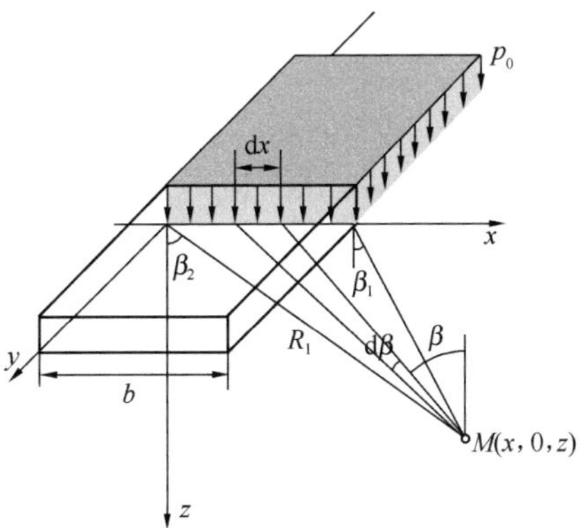

Figure 4.17 Additional stress under uniformly distributed strip load

It must be noted that the above calculation equation for additional stress in the soil is derived under the assumption that the soil is homogeneous. For the case of heterogeneous soil such as multilayer soil, anisotropic soil and soil with increasing deformation modulus with depth, it is necessary to refer to existing relevant literature for specific analysis.

必须注意，上述土中附加应力的计算式是在假设地基土均匀的情况下推导出来的。对于非均质土，如多层土、各向异性土和变形模量随深度增大的土，有必要参考现有的相关文献具体分析。

4.5 Effective Stress in Soil

4.5 土中有效应力

A soil can be visualized as a skeleton of solid particles enclosing continuous voids which contain water and/or air. On the one hand, for the range of stresses usually encountered in practice, the individual solid particles and water can be considered incompressible; air, on the other hand, is highly compressible. The volume of the soil skeleton as a whole can change due to rearrangement of the soil particles into new positions, mainly by rolling and sliding, with a corresponding change in the forces acting between particles. The actual compressibility of the soil skeleton will depend on the structural arrangement of the solid particles, i. e. the void ratio. In a fully saturated soil, since water is considered to be incompressible, a reduction in volume is possible only if some of the water can escape from the voids. In a dry or a partially saturated soil a reduction in volume is always possible, due to compression of the air in the voids, provided there is scope for particle

土体可以被视为固体颗粒的骨架，包裹着含有水或空气的连续孔隙。一方面对实践中通常遇到的应力范围，单个固体颗粒和水可以被认为是不可压缩的；另一方面，空气是高度可压缩的。土体颗粒主要通过滚动和滑动重新排列，引起土体骨架的整体体积变化，颗粒之间的作用力也会相应变化。土体骨架的实际压缩性将取决于固体颗粒的结构排列，即孔隙比。在完全饱和的土中，由于水被认为是不可压缩的，只有当一些水可以从孔隙中逸出时，体积才有可能减小。在干燥或部分饱和的土中，由于孔隙中空气可压缩，只要有颗粒重新排列的空间（即土体不处于最

rearrangement (i. e. the soil is not already in its densest possible state, $e>e_{min}$).

密实的状态，$e>e_{min}$），体积就可能减小。

Shear stress can be resisted only by the skeleton of solid particles, by means of reaction forces developed at the interparticle contacts. Normal stress may similarly be resisted by the soil skeleton through an increase in the interparticle forces. If the soil is fully saturated, the water filling the voids can also withstand normal stress by an increase in pore water pressure.

剪切应力只能由固体颗粒的骨架通过颗粒间接触处产生的反作用力来抵抗。土体骨架也可能通过增大颗粒间力来抵抗法向应力。如果土完全饱和，填充孔隙的水也可以通过增大的孔隙水压力来抵抗法向应力。

In 1943, Terzaghi recognized the importance of the forces transmitted through the soil skeleton from particle to particle and presented his Principle of Effective Stress, an intuitive relationship based on experimental data.

1943 年，太沙基认识到通过土体骨架颗粒间传递力的重要性，并提出了有效应力原理，如式（4.37）或式（4.38）所示，这是基于试验数据的直观关系。

$$\sigma=\sigma'+u \tag{4.37}$$

or

$$\sigma'=\sigma-u \tag{4.38}$$

The above equation is mathematical description of Principle of Effective Stress. The principle applies only to fully saturated soils, and relates the following three stresses:

(1) the total normal stress (σ) on a plane within the soil mass, being the force per unit area transmitted in a normal direction across the plane, imagining the soil to be a solid (single-phase) material;

(2) the pore water pressure (u), being the pressure of the water filling the void space between the solid particles;

(3) the effective normal stress (σ') on the plane, representing the stress transmitted through the soil skeleton only (i. e. interparticle forces).

上述方程是对有效应力原理的数学描述。该原理仅适用于完全饱和的土，并涉及以下三种应力：

（1）土体内某一平面上的总法向应力（σ），即单位面积在平面上沿法向传递的力，将土想象成固体（单相）材料；

（2）孔隙水压力（u）是填充固体颗粒之间的孔隙空间的水压力；

（3）平面上的有效法向应力（σ'），表示仅通过土骨架传递的应力（即粒间力）。

The principle can be represented by the following physical model. Consider a "plane" XX in a fully saturated soil, passing through points of interparticle contact only, as shown in Figure 4.18.

该原理可以用以下物理模型来表示。设想在完全饱和的土中有一个“平面”XX，该平面穿过颗粒间接触点，如图 4.18 所示。

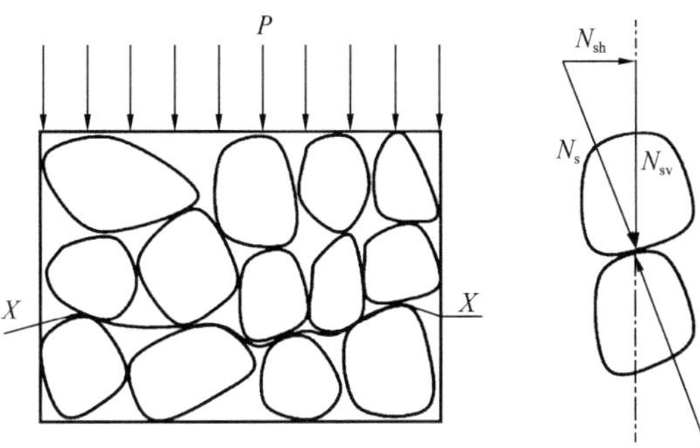

Figure 4.18 Stress transmission in saturated soil

The wavy plane XX is really indistinguishable from a true plane on the mass scale due to the relatively small size of individual soil particles. A normal force P applied over an area A may be resisted partly by interparticle forces and partly by the pressure in the pore water. The interparticle forces are very random in both magnitude and direction throughout the soil mass, but at every point of contact on the wavy plane may be split into components normal and tangential to the direction of the true plane to which XX approximates; the normal and tangential components are N_{sv} and N_{sh}, respectively. Then, the effective normal stress is approximated as the sum of all the components N_{sv} within the area A, divided by the area A, i. e.

从宏观尺度上看，由于单个土颗粒的尺寸相对较小，波浪形平面 XX 在质量尺度上与真实平面实际上难以区分。施加在面积 A 上的法向力 P 可以部分由粒间力抵抗，部分由孔隙水压力抵抗。在整个土体中，粒间力大小、方向都是随机的，但在波浪形平面上的每个接触点，都可以被分成垂直于真实平面（近似 XX 平面）法线方向和切线方向的分量；法向分量和切向分量分别为 N_{sv} 和 N_{sh}。然后，有效法向应力近似为面积 A 内所有分量 N_{sv} 的总和除以面积 A，即：

$$\sigma' = \frac{\sum N_{sv}}{A}$$

The total normal stress is given by

总法向应力为：

$$\sigma = \frac{P}{A}$$

If point contact is assumed between the particles, the pore water pressure will act on the plane over the entire area A. Then, for equilibrium in the direction normal to XX

如果假设颗粒之间存在点接触，则孔隙水压力将作用在整个平面 A 上。然后，在垂直于 XX 的方向上达到平衡。

$$P = \sum N_{sv} + uA$$

or

或

$$\frac{P}{A}=\frac{\sum N_{sv}}{A}+u$$

Therefore,

因此，

$$\sigma=\sigma'+u$$

The pore water pressure which acts equally in every direction will act on the entire surface of any particle, but is assumed not to change the volume of the particle (i. e. the soil particles themselves are incompressible); also, the pore water pressure does not cause particles to be pressed together. The error involved in assuming point contact between particles is negligible in soils, the actual contact area a normally being between 1% and 3% of the cross-sectional area A. It should be understood that σ' does not represent the true contact stress between two particles, which would be the random but very much higher stress N_{sv}/a.

在每个方向上均匀作用的孔隙水压力将作用于任意颗粒的整个表面，假设颗粒体积不变（即土颗粒本身是不可压缩的），且孔隙水压力不会导致颗粒被压在一起。在土体中，假设颗粒之间的点接触所涉及的误差可以忽略不计，实际接触面积 a 通常在横截面积 A 的 1%～3%之间。应该知道 σ' 并不代表两个颗粒之间的真实接触应力，真实接触压力将是随机的，但远高于 N_{sv}/a。

It should be noted that although pore water pressure cannot cause compression and mutual displacement between particles, according to Eq. (4.38), variations in pore water pressure can result in changes in effective stress, leading to interactions between particles, manifested macroscopically as deformation of the subsoil and changes in soil shear strength. Therefore, when the groundwater level changes, the effective stress in the saturated soil will change, resulting in additional stress in the subsoil. This additional stress will cause deformation of the soil. In engineering, special attention should be paid to changes in groundwater levels.

应注意：尽管孔隙水压力不会导致颗粒压缩和移动，但根据式（4.38），孔隙水压力的变化会导致有效应力变化，进而引起颗粒之间相互作用，宏观上表现为地基土变形和土体抗剪强度的变化。因此，当地下水位变化时，饱和土中的有效应力将随之变化，引起地基土的附加应力。这种附加应力将导致土体变形。在工程中，应特别注意地下水位的变化。

Problems

4.1 Given that the stress at a point in a pressurized soil layer is $\sigma_x=150$kPa, $\tau_{xy}=80$kPa, $\tau_{zx}=100$kPa. The measured pressure head height at this point is 2.0m. Calculate the effective major stress and minor at this point.

4.2 The independent foundation of a certain structure is shown in the Figure 4.19. A vertical concentrated load $F=320$kN with an eccentricity (length direction) of 0.5m is applied at the design ground elevation. The foundation is buried at a depth of $d=2$m, and the soil weight of the foundation is $\gamma=15.0\text{kN/m}^3$. The dimensions of the foundation bottom are: width × length$=b\times l=2\text{m}\times4\text{m}$. Calculate and plot the distribution map of substrate pressure. Assuming the additional stress coefficient at 3m below the foundation center point is 0. 134, what is the additional stress at 3m below the foundation center point?

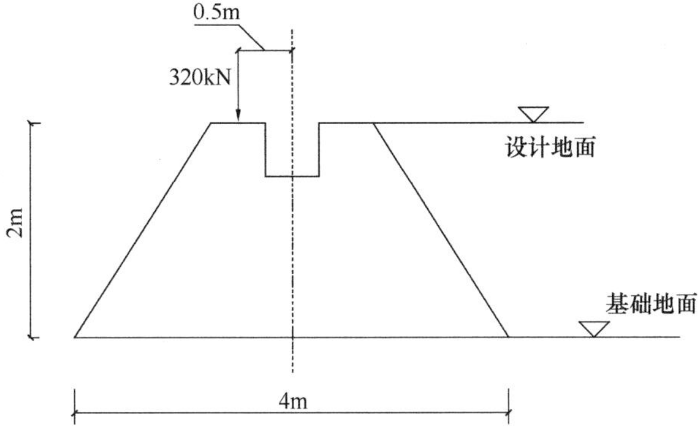

Figure 4. 19 Figure of problem 4. 2

4. 3 The independent foundation of a certain structure is shown in the Figure 4. 20, with $F=3600$kN, a foundation burial depth of $d=2.0$m, a foundation soil weight of $\gamma=20.0\text{kN/m}^3$, and a foundation bottom size of length × width $=l\times b=10\text{m}\times 6\text{m}$. Calculate the distribution of base pressure. Assuming the additional stress coefficient of the rectangular distributed load at 2m below the center point of the foundation is 0. 198, calculate the additional stress at 2m below the center point of the foundation.

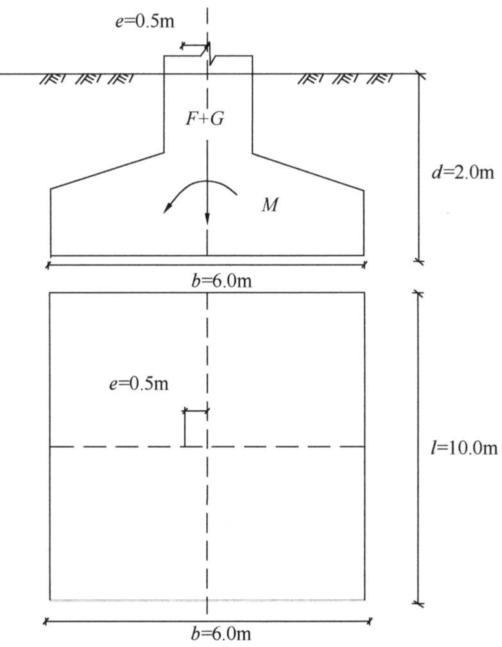

Figure 4. 20 Figure of problem 4. 3

Chapter 5 Compressibility of Soil

第 5 章 土的压缩性

5.1 Introduction

A stress increase caused by the construction of foundations or other loads compresses soil layers. The compression is caused by (1) deformation of soil particles, (2) relocations of soil particles, (3) expulsion of water or air from the void spaces. In general, the soil settlement caused by loads may be divided into three broad categories:

(1) Elastic settlement (or immediate settlement), which is caused by the elastic deformation of dry soil and moist and saturated soils without any change in the moisture content. Elastic settlement calculations generally are based on equations derived from the theory of elasticity.

(2) Primary consolidation settlement, which is the result of a volume change in saturated cohesive soils because of expulsion of the water that occupies the void spaces.

(3) Secondary consolidation settlement, which is observed in saturated cohesive soils and organic soil and is the result of the plastic adjustment of soil fabrics. It is an additional form of compression that occurs at constant effective stress.

The total settlement of a foundation can then be given as:

5.1 概述

由地基或其他荷载引起的应力增大会导致土层的压缩。这种压缩主要由以下三方面引起：(1) 土体颗粒的变形；(2) 土体颗粒的位移；(3) 孔隙中的水或空气被排出。通常，由荷载引起的土体沉降可分为三类：

(1) 弹性沉降（或瞬时沉降），是由干土、湿土和饱和土在含水量不变的情况下发生弹性变形所引起的。弹性沉降计算通常基于弹性理论推导的方程。

(2) 主固结沉降，是饱和黏性土中孔隙水被排出导致体积变化的结果。

(3) 次固结沉降，常见于饱和黏性土和有机土中，是土体结构塑性调整的结果。是在恒定有效应力下发生的另一种压缩形式。

基础的总沉降量的表达式如式（5.1）所示。

$$s = s_c + s_s + s_e \tag{5.1}$$

where s is total settlement; s_c is primary consolidation settlement; s_s is secondary consolidation settlement; s_e is elastic settlement, it can be derived from Boussinesq's Fundamental Solution.

式中，s 为总沉降量；s_c 为主固结沉降量；s_s 为次固结沉降量；s_e 为弹性沉降量，可根据布辛奈斯克基本解推导得出。

$$s_e = \omega_0 \frac{(1-\mu^2)}{E} b p_0 \tag{5.2}$$

where ω_0 is the settlement influence coefficient of the center point; μ and E are the Poisson's ratio and elastic

式中，ω_0 为中心点的沉降影响系数；μ 和 E 分别为地基土的泊

modulus of subsoil, respectively; b is the width of rectangular foundation or radius of circular foundation; p_0 is the additional contact pressure under foundation base.

松比和弹性模量；b 为矩形基础宽度或圆形基础半径；p_0 为基础底部的附加接触压力。

The sum of the first two is referred to as consolidation settlement, and its calculation method will be presented in Chapter 6.

前两项之和被称为固结沉降，其计算方法将在第 6 章中介绍。

When foundations are constructed on very compressible clays, the consolidation settlement can be several times greater than the elastic settlement.

当基础建在可压缩性较强的黏土上时，固结沉降可能为弹性沉降的几倍。

This chapter will cover the following: (1) consolidation test and compressibility index; (2) the influence of stress history on compressibility; (3) deformation modulus of soil.

本章将介绍以下内容：(1) 固结试验和压缩性指标；(2) 应力历史对压缩性的影响；(3) 土的变形模量。

5.2 Consolidation Test and Compressibility Index

5.2 固结试验和压缩性指标

5.2.1 Consolidation Test

5.2.1 固结试验

The characteristics of a soil during one-dimensional consolidation can be determined by means of the oedometer test. Figure 5.1 shows diagrammatically a cross-section through an oedometer. The test sample is in the form of a disc of soil, held inside a metal ring and lying between two porous stones. The upper porous stone, which can move inside the ring with a small clearance, is fixed below a metal loading cap through which pressure can be applied to the sample. The whole assembly sits in an open cell of water to which the pore water in the sample has free access. The ring confining the sample may be either fixed (clamped to the body of the cell) or floating (being free to move vertically); the inside of the ring should have a smooth polished surface to reduce side friction. The confining ring imposes a condition of zero lateral strain on the sample. The compression of the sample under pressure is measured by means of a dial gauge or electronic displacement transducer operating on the loading cap.

土体在一维固结过程中的特性可通过固结试验确定。图 5.1 为固结仪及其横断面，其中试样被固定在金属环内，位于两块透水石之间，形状呈土盘状。上部透水石在环内可略微移动，其被固定在金属加载帽下方，通过该加载帽可向试样施加压力。整个装置位于开放式水容器中，试样中的孔隙水可自由进入，装置中约束试样的金属环是固定的（夹在水容器上）或浮动的（可竖向自由移动），环内侧有光滑的抛光面以减小侧面摩擦，同时金属环对试样施加的侧向应变为零，试样在各压力下的压缩量通过安装在加载帽上的千分表或电子位移传感器测量。

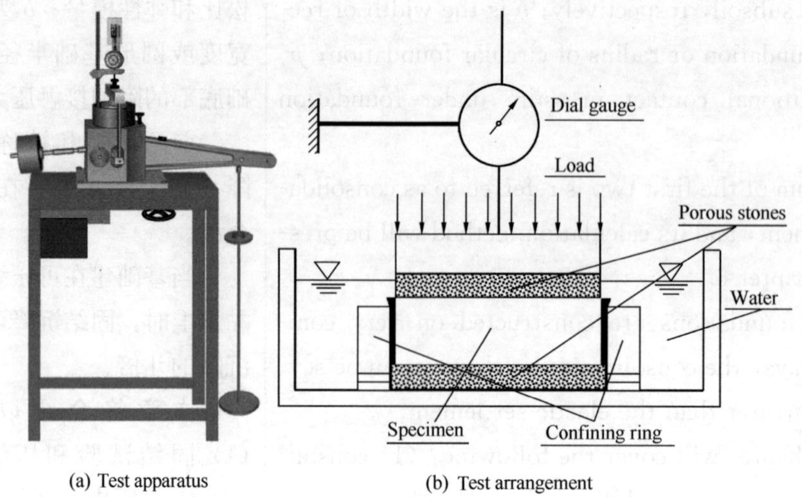

(a) Test apparatus (b) Test arrangement

Figure 5.1 The oedometer

The test procedure has been standardized in " Standard for Geotechnical Testing Methods" GB/T 50123—2019, which specifies that the oedometer shall be of the fixed ring type. The initial pressure (total stress) applied will depend on the type of soil; following this, a sequence of pressures is applied to the specimen, each being double the previous value. Each pressure is normally maintained for a period of 24h (in exceptional cases a period of 48h may be required), compression readings being observed at suitable intervals during this period. At the end of the increment period, when the excess pore water pressure has completely dissipated, the applied total stress equals the effective vertical stress in the sample. The results are presented by plotting the thickness (or percentage change in thickness) of the sample or the void ratio at the end of each increment period against the corresponding effective stress. The effective stress may be plotted to either a natural or a logarithmic scale, though the latter is normally adopted due to the reduction in volume change in a given increment as total stress increases. If desired, the expansion of the sample can additionally be measured under successive decreases in applied pressure to observe the swelling behavior. However, even if the swelling characteristics of the soil are not required, the

固结试验过程已在《土工试验方法标准》GB/T 50123—2019 中标准化：规定固结仪应为固定环型，所施加的初始压力值（总应力）取决于土体类型，此后对试样施加逐级增大的压力，每级压力为前一级压力的两倍，通常各个压力值应维持 24h（在特殊情况下可能需要 48h），在此期间每隔适当时间观察其压缩量。每个压力值对应的压缩增量结束，当超孔隙水压力完全消散后，所施加总应力等于试样中有效竖向应力，通过每个压缩增量结束时试样的厚度（或厚度变化的百分比）或孔隙比与相应有效应力的关系，可以呈现试验结果。其中有效应力可按自然刻度或对数刻度绘制，但由于总应力增加时体积变化减小，通常采用后者。若分析需要，可在连续降低施加压力的情况下测量试样的回弹量并观察其回弹情况，即使不需要观察土体的回弹特性，也应测量最终压力解除后试样的回

expansion of the specimen due to the removal of the final pressure should be measured.

弹情况。

China code recommends that a minimum of two tests are conducted in a given soil stratum. This value should be doubled if there is considerable discrepancy in the measured compressibility, especially if there is little or no previous experience relating to the soil in question.

我国规范建议对特定土层至少进行两次试验，若所测压缩性存在显著差异，特别缺乏所测土体相关试验经验，试验次数应加倍。

The void ratio at the end of each increment period can be calculated from the displacement readings and either the water content or the dry weight of the specimen at the end of the test. Referring to the phase diagram in Figure 5.2, the method of calculation is as follows:

根据位移读数和试验结束时试样的含水量或干重，可计算各个压力下压缩增量结束时试样孔隙比。如图 5.2 所示，计算方法如下：

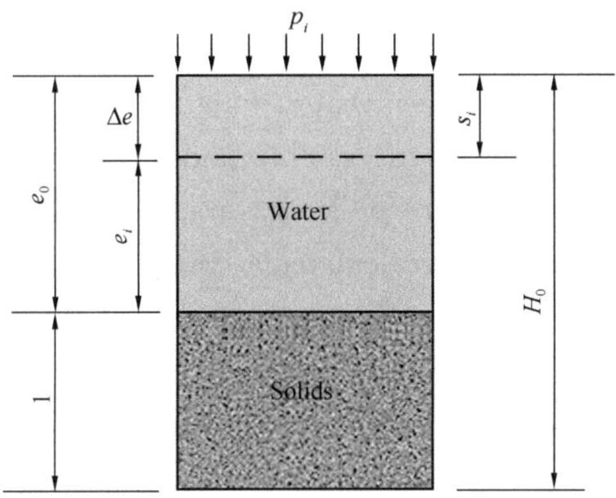

Figure 5.2 Phase diagram

Assuming that the volume of soil particles remains unchanged and the cross-sectional area of the sample remains unchanged before and after compression. Therefore, the volume of the sample before compression is

假设土颗粒体积不变，试样在受压前后的横截面面积保持不变，则压缩前试样的体积可表示为式（5.3）。

$$V_0 = 1 + e_0 \tag{5.3}$$

The volume of the sample after compression stabilization under pressure p_i is

试样在压力 p_i 下压缩稳定后的体积可用式（5.4）表示。

$$V_i = 1 + e_i \tag{5.4}$$

where V_0 and V_i are respectively volume of the sample before and after compression; e_0 is initial void ratio of the sample and e_i is void ratio of the sample under the action of p_i after compression.

式中，V_0 和 V_i 分别为试样压缩前后的体积；e_0 为试样的初始孔隙比；e_i 为试样在 p_i 作用下压缩后的孔隙比。

The cross-sectional area of the sample before and after compression are

试样在压缩前后的横截面面积如式（5.5）所示。

$$A_0 = \frac{1+e_0}{H_0},\ A_i = \frac{1+e_i}{H_i} \tag{5.5}$$

where A_0 and A_i are respectively cross-sectional area of the sample before and after compression; H_0 is initial height of the sample and H_i is height of the sample after compression.

式中，A_0 和 A_i 分别为试样压缩前后的横截面面积；H_0 为试样的初始高度；H_i 为试样压缩后的高度。

Therefore, from equation (5.5), it can be derived that

因此，由式（5.5）可推导出式（5.6）。

$$e_i = e_0 - \frac{s_i}{H_0}(1+e_0) \tag{5.6}$$

where s_i is the compression amount of the sample when it is compressed to stability under the action of p_i, $s_i = H_0 - H_i$.

式中，s_i 为试样在 p_i 作用下被压缩至稳定时的压缩量，$s_i = H_0 - H_i$。

5.2.2 Compressibility Coefficient and Compressibility Index

5.2.2 压缩系数和压缩指数

From the above test, it can be seen that as long as the stable compression amount s_i of the sample under the action of various levels of pressure p_i is measured, the corresponding void ratio e_i can be calculated according to Eq. (5.6). If the pressure p_i is taken as the horizontal axis and the corresponding void ratio is taken as the vertical axis, it can be plotted the e-p curve as shown in Figure 5.3 (a). If the horizontal axis is changed to the logarithmic axis, the e-lg$p$ curve can be obtained as shown in Figure 5.3 (b).

从上述试验可知，只要测定试样在各级压力 p_i 作用下的稳定压缩量 s_i，就可以按式（5.6）计算相应的孔隙比 e_i。若以压力 p_i 为横坐标，以相应的孔隙比为纵坐标，则可绘制出如图 5.3（a）所示的 e-p 曲线。如果将横坐标改为对数坐标，可得如图 5.3(b) 所示的 e-lg$p$ 曲线。

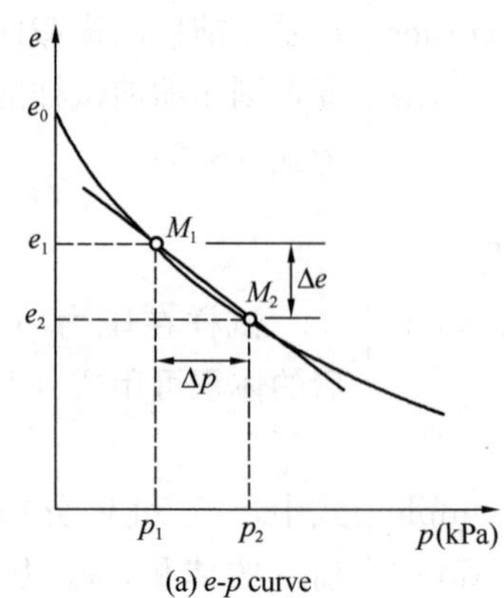

(a) e-p curve

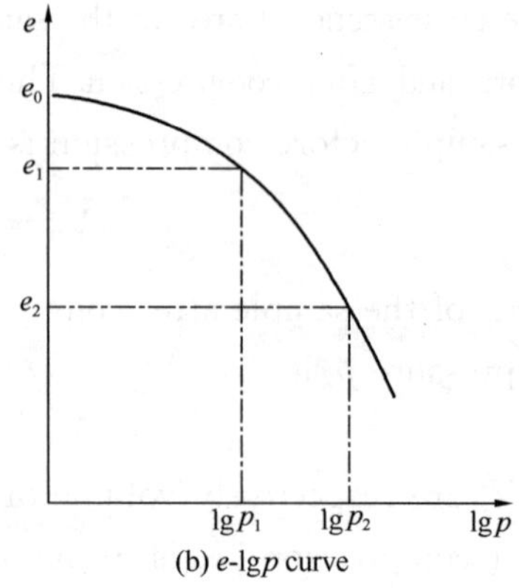

(b) e-lgp curve

Figure 5.3 Compression curve

The compressibility coefficient of soil defined as the ratio (MPa^{-1}) of the decrease in void ratio to the increase in effective compressive stress under lateral confinement conditions, which is the slope of the tangent line in a certain pressure segment of the e-p curve. The pressure section in the subsoil should be taken from the self-weight stress of the soil to the sum of the self-weight stress and additional stress of the soil. The steeper the curve, the more significant the decrease in soil void within the same pressure range, resulting in higher soil compressibility. So, the slope a of the tangent at any point on the curve represents the compressibility of the soil under the corresponding pressure p.

土体的压缩系数是指土体在侧限条件下，孔隙比减小量与有效压应力增量的比值（MPa^{-1}），即 e-p 曲线上某压力段的切线斜率。对于地基土而言，土体压力段应从土体自重应力开始，到土体自重应力与附加应力之和为止。其曲线越陡，表明在相同压力范围内土体孔隙的减少越显著，土体压缩性越高，因此，曲线上任意点的切线斜率 a 表示为在相应压力 p 下的土体压缩性，表达式见式（5.7）。

$$a = -\frac{\mathrm{d}e}{\mathrm{d}p} \tag{5.7}$$

The negative sign in the equation indicates that as the pressure p increases, the void ratio e gradually decreases. In practice, the compressibility characterized by the increase of the original pressure p_1 in soil at a certain point to the total pressure p_2 after external loading is generally focused on. As shown in Figure 5.3 (a), assuming the pressure increases from p_1 to p_2 and the corresponding void ratio decreases from e_1 to e_2, the void ratio changes corresponding to the pressure increment $\Delta p = p_2 - p_1$ is $\Delta e = e_1 - e_2$. At this point, the compressibility of the soil can be represented by the slope of the secant line $M_1 M_2$ in the graph. Then,

式中负号表示随压力 p 的增加，孔隙比 e 逐渐减小。在实际应用中，通常会重点考虑土体在受到外部荷载后，某点由初始压力 p_1 增大至总压力 p_2 时表现出的压缩性特征。如图 5.3（a）所示，假设压力从 p_1 增大至 p_2，相应的孔隙比从 e_1 减小为 e_2，则压力增量 $\Delta p = p_2 - p_1$ 对应的孔隙比变为 $\Delta e = e_1 - e_2$。此时，图中割线 M_1M_2 的斜率可表示土体的压缩性，用式（5.8）表示。

$$a = \frac{\Delta e}{\Delta p} = \frac{e_1 - e_2}{p_2 - p_1} \tag{5.8}$$

where a is the compressibility coefficient of soil (MPa^{-1}); p_1 is the vertical self-weight stress (MPa) in the soil at a certain depth of the subsoil refers to the "original pressure" at a certain point in the soil. p_2 is the sum of the (vertical) self-weight stress and (vertical) additional stress in the soil at a certain depth of the subsoil (MPa) refers to the " total stress" at a certain point in the soil; e_1 and e_2 are the void ratio after compression stabilization under the action of p_1 and p_2, respectively.

式中，a 为土体压缩系数（MPa^{-1}）；p_1 为某深度处的地基土垂直自重应力（MPa），代表土中某点的“原始压力”；p_2 指某深度处地基土的（垂直）自重应力与（垂直）附加应力之和（MPa），代表土中某点的“总应力”；e_1 和 e_2 分别为在 p_1 和 p_2 的作用下压缩稳定后的孔隙比。

For ease of comparison, the compressibility of soil is usually evaluated using the compression coefficient $a_{1\text{-}2}$ when the pressure range increases from $p_1 = 0.1\text{MPa}$ (100kPa) to $p_2 = 0.2\text{MPa}$ (200kPa), as follows:

When $a_{1\text{-}2} < 0.1\text{MPa}^{-1}$, it is a low compressibility soil;

When $0.1\text{MPa}^{-1} \leqslant a_{1\text{-}2} < 0.5\text{MPa}^{-1}$, it is a moderately compressible soil;

When $a_{1\text{-}2} \geqslant 0.5\text{MPa}^{-1}$, it is a highly compressible soil.

为便于比较，通常使用压力从 $p_1=0.1\text{MPa}$（100kPa）增大至 $p_2=0.2\text{MPa}$（200kPa）时对应的压缩系数 $a_{1\text{-}2}$ 评估土体压缩性：

当 $a_{1\text{-}2} < 0.1\text{MPa}^{-1}$ 时，土体为低压缩性土；

当 $0.1\text{MPa}^{-1} \leqslant a_{1\text{-}2} < 0.5\text{MPa}^{-1}$ 时，土体为中等压缩性土；

当 $a_{1\text{-}2} \geqslant 0.5\text{MPa}^{-1}$ 时，土体为高压缩性土。

The compressibility index of soil defined as the ratio of the decrease in porosity of the soil under lateral confinement conditions to the commonly used logarithmic increment of effective compressive stress, that is, the slope of a straight line in a certain pressure segment of the e-lg$p$ curve. When the e-p curve of soil is plotted as a semi logarithmic e-lg$p$ curve, its posterior pressure segment approaches a straight line (Figure 5.3b), and its slope C_c is

土体压缩指数指土在侧限条件下受压时，孔隙率的减小量与有效压应力对数增量的比值，即 e-lg$p$ 曲线某压力段中的直线斜率。当土体的 e-p 曲线绘制为半对数 e-lg$p$ 曲线时，其后压力段接近一条直线（图 5.3b），其斜率 C_c 即为土体的压缩指数，可用式（5.9）表示。

$$C_c = \frac{e_1 - e_2}{\lg p_2 - \lg p_1} = \frac{\Delta e}{\lg(p_2/p_1)} \tag{5.9}$$

where C_c is the compressibility index of soil.

Like the compressibility coefficient a, the larger the compressibility index C_c value, the higher the compressibility of the soil. The C_c value of low compressibility soil is generally less than 0.2, while a C_c value greater than 0.4 indicates high compressibility soil.

式中，C_c为土体压缩指数。

与压缩系数 a 相同，压缩指数 C_c值越大，土体压缩性越强。低压缩性土的 C_c 值一般小于 0.2，而 C_c值大于 0.4 则表明土体具有高压缩性。

5.2.3 Compression Modulus and Coefficient of Volume Compressibility of Soil

5.2.3 压缩模量和体积压缩系数

The definition of the compression modulus E_s of soil is the ratio of the vertical additional compressive stress to the vertical strain of the soil under lateral confinement conditions (MPa). It is the second compressibility index obtained from the e-p curve. If the change in void ratio of the sample in the e-p curve, $\Delta e = e_1 - e_2$, is known, the corresponding change in sample height, $\Delta H = H_1 - H_2$, can be calculated backwards.

土体压缩模量 E_s 为在侧限条件下，土体的竖向附加压应力与土体竖向应变之比（MPa），其是从 e-p 曲线中获得的第二个压缩性指标。如果已知 e-p 曲线中试样孔隙比的变化 $\Delta e = e_1 - e_2$，则可反向计算出相应的试样高度变化 $\Delta H = H_1 - H_2$。

Under the condition that the pressure increment $\Delta p = p_2 - p_1$ remains unchanged before and after the application of pressure increment $\Delta p = p_2 - p_1$, and assuming it is equal to 1, the soil particle height $H_1/(1+e_1)$ under compression p_1 is equal to the soil particle height $H_2/(1+e_2)$ under compression p_2. Therefore,

在施加压力增量 $\Delta p = p_2 - p_1$ 不变的条件下，假设 Δp 等于 1，则在压力 p_1 作用下的土体高度 $H_1/(1+e_1)$ 等于在压力 p_2 作用下的土体高度 $H_2/(1+e_2)$，表达式为式（5.10）或式（5.11）。

$$\frac{H_2}{H_1} = \frac{1+e_2}{1+e_1} \tag{5.10}$$

or

$$\varepsilon_v = \frac{\Delta H}{H_1} = \frac{e_1 - e_2}{1+e_1} \tag{5.11}$$

where ε_v is the vertical strain of sample.
Since $\Delta e = a\Delta p$,

式中，ε_v为试样的竖向应变。
由于 $\Delta e = a\Delta p$，可得式（5.12）：

$$E_s = \frac{\Delta p}{\varepsilon_v} = \Delta p / \left(\frac{e_1 - e_2}{1+e_1}\right) = \frac{\Delta p(1+e_1)}{e_1 - e_2} = \frac{1+e_1}{a} \tag{5.12}$$

The above equation represents that under lateral confinement conditions, when the stress in the soil does not change significantly, the increment of compressive stress is proportional to the increment of compressive strain, and the proportionality coefficient E_s is called the sample modulus of the soil, or the lateral confinement modulus. The smaller the sample modulus E_s value of soil, the higher its compressibility.

上式表示，在侧限条件下，当土体中应力没有发生显著变化时，则压应力增量与压应变增量成正比，比例系数 E_s 被称为土体的压缩模量，也称为侧限模量。土体压缩模量 E_s 值越小，其压缩性越好。

The coefficient of volume compressibility (m_v) of soil is the third compressibility index obtained from the e-p curve. Its definition is the ratio of the vertical (volume) strain to the vertical additional compressive stress of the soil under lateral confinement conditions (MPa^{-1}), also known as the unidirectional volume compressibility coefficient, which is the reciprocal of the soil's sample modulus.

土体的体积压缩系数（m_v）是从 e-p 曲线获得的第三个压缩系数。其是在侧限条件下，土体的竖向（体积）应变与竖向附加压应力之比（MPa^{-1}），也称为单向体积压缩系数，是土体压缩模量的倒数。其中体积压缩系数可表达为式（5.13）。

$$m_v = \frac{1}{E_s} = \frac{a}{1+e_1} \tag{5.13}$$

Like the compressibility coefficient and compressibility index, the larger the volume compressibility coefficient m_v value, the higher the compressibility of the soil.

与压缩系数和压缩指数相同，体积压缩系数 m_v 值越大，土体的压缩性越高。

During the indoor consolidation test, if the pressure is increased to a certain value p_i and no longer applied (corresponding to the compression curve of segment ab in the e-p curve shown in Figure 5.4), but gradually reduced to zero, the rebound of the sample can be observed.

在室内固结试验中，如果将压力增大到一定值 p_i 后停止施加（对应于图 5.4 所示的 e-p 曲线中 ab 段的压缩曲线），然后将压力值逐渐减小至零，可以观察试样的回弹现象。

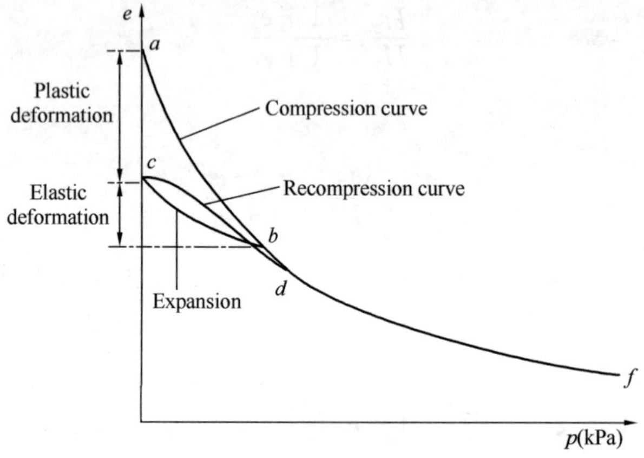

Figure 5.4 Rebound curve and recompression curve of soil

The void ratio of the sample after rebound stabilization under various levels of pressure can be measured, and the corresponding relationship curve between void ratio and pressure can be drawn. The bc curve in Figure 5.4 is called the rebound curve. Due to the compression deformation of the soil sample under pressure p_i, after unloading, the sample cannot fully recover to point a of the initial void ratio e_0. This indicates that the excavated compression deformation is composed of elastic deformation and residual deformation, with the latter being the main component. If the pressure is gradually increased again, the void ratio of the specimen can be measured after being compressed and stabilized at each level of pressure, and a recompression curve can be drawn, as shown in the cdf in the figure. The df segment seems to be a continuation of the ab segment, as if it has not undergone a process of pressure relief and repressurization.

测量试样在不同压力水平下回弹稳定后的孔隙比，并绘制孔隙比与压力之间的关系曲线。图 5.4 中的 bc 曲线被称为回弹曲线，由于试样在压力 p_i 下会产生压缩变形，卸载后无法完全恢复到初始孔隙比 e_0 的点 a。表明压缩变形由弹性变形和残余变形组成，其中残余变形为主要部分。如果再次逐渐增大压力，则可以测量试样在各个压力水平下被压缩并稳定后的孔隙比，并绘制出再压缩曲线，如图中的 cdf 段所示。df 段似乎是 ab 段的延续，仿佛没有经历减压和再增压的过程。

By using the e-p curves of compression, rebound, and recompression, it is possible to analyze large area

通过使用压缩、回弹和再压缩的 e-p 曲线，可以分析某些

and deep buried foundation. After excavation, the subsoil experiences significant pressure reduction (stress relief), resulting in bottom rebound. Therefore, when estimating foundation settlement, the rebound of the excavated foundation soil should be considered, and a rebound and compression test of the soil should be conducted. The pressure applied should be consistent with the actual loading, unloading, and reloading conditions at a certain point in the subsoil. To calculate the rebound deformation of the soil at the bottom of the excavation pit, the rebound modulus E_c of the subsoil must be determined from the e-p curve of the rebound and recompression of the consolidation test. The definition of the rebound modulus of subsoil is the ratio (MPa) of the vertical additional compressive stress to the vertical strain of the soil when unloaded or reloaded under lateral limiting conditions. It is the fourth compressibility index obtained based on the e-p curve, as detailed in the "Code for Design of Building Foundation" GB 50007—2011. By using the e-lg$p$ curves of compression, rebound, and recompression, the influence of stress history on the compressibility of soil can also be analyzed.

具有大面积和深埋的基础。开挖后，地基土的压力显著降低（应力消除），导致底部回弹。因此，在估算基础沉降时，应考虑开挖后地基土的回弹，并进行土体的回弹和压缩试验。所施加的压力应与地基土中某一点的实际加载、卸载和再加载条件一致。为计算基坑坑底回弹变形，必须根据固结试验的回弹和再压缩 e-p 曲线，确定地基土的回弹模量 E_c。地基土回弹模量为在侧限条件下卸载或再加载时，土体竖向附加压应力与竖向应变的比值（MPa）。这是根据 e-p 曲线获得的第四个压缩性指标，详细规定见《建筑地基基础设计规范》GB 50007—2011。此外，通过使用压缩、回弹和再压缩的 e-lg$p$ 曲线，还可以分析应力历史对土体压缩性的影响。

5.3 The Influence of Stress History on Compressibility

The natural soil layer has undergone the maximum consolidation pressure in history (referring to the maximum vertical effective stress experienced by the soil during the consolidation process), which is called pre-consolidation pressure, or pre-consolidation pressure. According to the stress history, soil (layer) can be divided into three categories: normally consolidated soil (layer), over consolidated soil (layer), and under consolidated soil (layer) . The pre-consolidation pressure experienced by normally consolidated soils in history is equal to the weight of the existing cover soil; over consolidated soils (OC soils) have historically experienced pre-consolidation pressures greater than the weight of the existing overlying soil; The pre-consolidation pressure

5.3 应力历史对压缩性的影响

天然土层经历过的最大固结压力（指土层在固结过程中所受的最大竖向有效应力）被称为先期固结压力或前期固结压力。根据应力历史，土（层）可以分为三类：正常固结土（层）、超固结土（层）和欠固结土（层）。正常固结土所经历的先期固结压力等于现有上覆土层的重量；超固结土（OC 土）所经历的先期固结压力大于现有上覆土层的重量；而欠固结土的先期固结压力则低于现有上覆土层的重量。在研究沉积土层的应力历史时，通

of under consolidated soils is lower than the weight of the existing cover soil. When studying the stress history of sedimentary soil layers, the ratio of pre-consolidation pressure to the weight of existing overlying soil is usually defined as Over Consolidation Ratio (OCR) as follows:

常将先期固结压力与现有上覆土层重量的比值定义为超固结比（OCR），其表达式为式（5.14）。

$$OCR = p_c / p_1 \tag{5.14}$$

where p_c is the pre-consolidation pressure (kPa); p_1 is the existing coverage self-weight stress (kPa).

式中，p_c为固结前压力（kPa）；p_1是现有覆盖层的自重应力（kPa）。

The over consolidation ratios of normally consolidated soil (layer), over consolidated soil (layer), and under consolidated soil (layer) are OCR=1, OCR>1, and OCR<1, respectively.

正常固结土（层）、超固结土（层）和欠固结土（层）的超固结比（OCR）分别为 OCR=1、OCR>1 和 OCR<1。

As shown in Figure 5.5, the A-type covering soil layer gradually deposited to the current surface. Due to a long geological period, it has reached a stable consolidation state under the self-weight of the soil (Figure 5.5a). Its initial consolidation pressure p_c is equal to the self-weight stress of the existing covering soil $p_1 = \gamma h$ (γ is the natural unit weight of homogeneous soil, and h is the depth of the calculation point below the current surface), so A-type soil is a normally consolidated soil. The B-type covering soil layer was originally a relatively thick sedimentary layer in history, and has reached a stable state under the self-weight of the soil. The dashed line in Figure 5.5 (b) represents the surface of the sedimentary layer at that time. Later, due to erosion caused by flowing water or glaciers, the current surface was formed. Therefore, the pre-consolidation pressure $p_c = \gamma h_c$ (where h_c is the calculated depth below the ground before erosion) exceeds the existing self-weight stress p_1 of the soil. Therefore, B-type soil is a super consolidated soil, and the larger its OCR value, the greater the super consolidation effect.

如图 5.5 所示，A 类覆盖土层逐渐沉积到当前地表。由于经历了漫长的地质时期，它已在土体自重下达到了稳定的固结状态（图 5.5a）。其初始固结压力 p_c 等于现有覆盖土层的自重应力 $p_1 = \gamma h$（γ 为均质土的自然重度，h 是计算点在当前地表以下的深度），因此 A 类土体是正常固结土。

B 类覆盖土层原本是较厚的沉积层，并在土体自重下达到了稳定状态。图 5.5(b) 中的虚线代表当时沉积层的表面。后来，由于流水或冰川的侵蚀作用，形成了当前地表，因此，其先期固结压力 $p_c = \gamma h_c$（h_c为侵蚀前地面以下的计算深度）超过了土体现有的自重应力 p_1。因此，B 类土体是超固结土，其 OCR 值越大，超固结效应越显著。

The C-type soil layer, like the A-type soil layer, gradually deposited to the current surface, but the difference is that it did not reach a consolidated stable state. For example, newly deposited cohesive soil, artificial fill, etc., due to the short period of time after

C 类土层与 A 类土层类似，逐渐沉积到当前地表，但不同之处在于其未达到固结稳定状态。例如，新沉积的黏性土、人工填

sediment-ation, their self-weight consolidation has not yet been completed. The dashed line in Figure 5.5 (c) represents the surface after consolidation in the future. Therefore, p_c (where $p_c = \gamma h_c$, h_c represents the calculation point depth below the ground after consolidation) is still less than the existing self-weight stress p_1 of the soil. Therefore, C-type soil is an under consolidated soil.

土等，由于沉积后时间较短，其自重固结尚未完成。图 5.5（c）中的虚线代表未来固结后的表面。因此，p_c（$p_c = \gamma h_c$，h_c 为固结后地面以下的计算点深度）仍然小于土体现有的自重应力 p_1。因此，C 类土体是欠固结土。

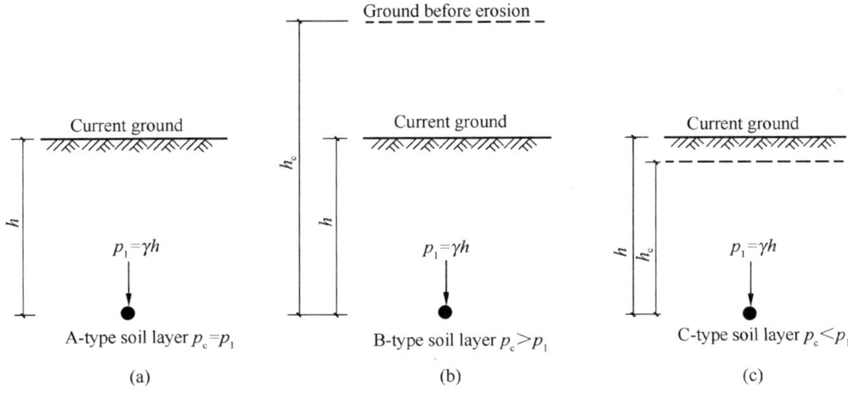

Figure 5.5 Sedimentary layers classified according to pre-consolidation pressure

When considering the stress history of soil layers for deformation calculations, high-pressure consolidation tests should be conducted to determine compressibility indicators such as pre-consolidation pressure and compression index. The test results are represented by e-lg$p$ curves. The most commonly used method for determining the pre-consolidation pressure p_c is the empirical plotting method proposed by A. Cassagrande at 1936, with the following plotting steps (Figure 5.6):

在分析土层的应力历史以计算变形时，应进行高压固结试验，以确定如先期固结压力和压缩指数等压缩性指标。试验结果通过 e-lg$p$ 曲线来表示。确定先期固结压力 p_c 最常用的方法是 A. Cassagrande 1936 年提出的经验作图法，其作图步骤如图 5.6 所示：

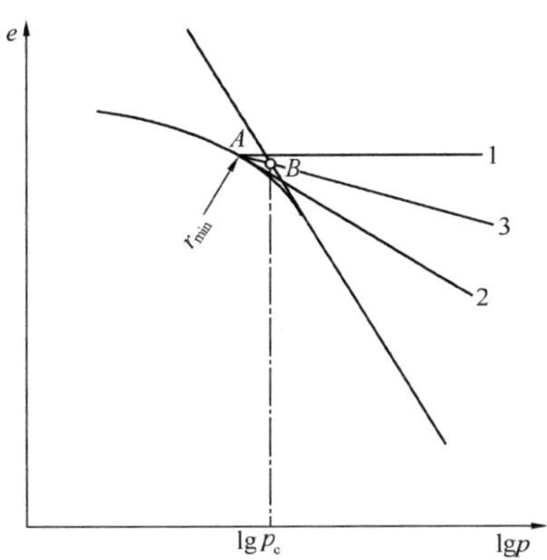

Figure 5.6 Method for determining pre-consolidation pressure

(1) Find the point A with the smallest curvature radius on the e-lg$p$ curve, and draw a horizontal line $A1$ and a tangent line $A2$ passing through point A;

(2) The bisector $A3$ of $\angle 1A2$ intersects with the extension of the straight line segment in the e-lg$p$ curve at point B;

(3) The effective stress corresponding to point B is the pre-consolidation pressure p_c.

It must be pointed out that using this simple empirical drawing method requires high quality for soil sampling, and appropriate scales should be selected when drawing the e-lg$p$ curve. Otherwise, it is sometimes difficult to find a sudden change in point A, and therefore, not all reliable results can be obtained.

(1) 在 e-lg$p$ 曲线上找到曲率半径最小的点 A，并过点 A 绘制一条水平线 $A1$ 和一条切线 $A2$；

(2) $\angle 1A2$ 的角平分线 $A3$ 与 e-lg$p$ 曲线上直线段的延长线相交于点 B；

(3) 与点 B 对应的有效应力即为先期固结压力 p_c。

必须指出，使用这种简单的经验作图法对取土质量要求较高，并且在绘制 e-lg$p$ 曲线时应选择合适的比例尺。否则，有时难以找到点 A 处的突变，从而无法得到可靠的结果。

5.4 Deformation Modulus of Soil

5.4.1 Shallow Plate Load Test and Deformation Modulus

Shallow plate loading test refers to field loading test, which is a basic in-situ testing of subsoil in engineering geological exploration work. Before the experiment, a load frame is erected in the on-site test pit to transfer the applied load to the formation through a pressure plate, in order to test the mechanical properties of the soil within the influence range of the additional stress on the shallow foundation, including measuring the deformation modulus of the soil, the bearing capacity of the foundation, and studying the soil's collapsibility properties. The pressure plate should have sufficient rigidity, and the bottom area of the pressure plate should not be less than 0.25m^2. For soft soil, it should not be less than 0.5m^2 (square side length 0.707m×0.707m or circular diameter 0.798m). To simulate local loads on the surface of a half space foundation, the width of the foundation pit should not be less than three times the width or diameter of the bearing plate, as specified in the "Code for Design of Building Foundation" GB 50007—2011.

5.4 土的变形模量

5.4.1 浅板载荷试验和变形模量

浅层平板载荷试验是指现场加载试验，是工程地质勘探工作中地基土的基本原位测试。试验前，在现场试验坑内架设一个载荷架，通过压力板将施加的荷载传递到地层，以测试浅基础附加应力影响范围内土的力学性能，包括测量土的变形模量、基础的承载力，并研究土的湿陷性。压板应具有足够的刚度，压板底部面积不应小于 0.25m^2。对于软土，不应小于 0.5m^2（方形边长 0.707m×0.707m 或圆形直径 0.798m）。根据《建筑地基基础设计规范》GB 50007—2011 的规定，为了模拟半空间基础表面的局部荷载，基坑的宽度不应小于承重板宽度或直径的 3 倍。

The load frame of the two types of jacks shown in Figure 5.7 consists of three parts: a load stabilizing device, a reaction device, and an observation device. The load stabilizing device includes a pressure plate, a column, a loading jack, and a stabilizer; Reaction devices include anchor systems or pile systems, etc.; The observation device includes a dial gauge and a fixed bracket. The load test points are usually arranged near the technical borehole where the sample is taken. When the geological structure is simple, the distance should not exceed 10m, and in other cases, it should not exceed 5m, but it should not be less than 2m. Attention must be paid to maintaining the original structure and natural humidity of the test soil layer. It is advisable to level the surface to be tested with coarse or medium sand layer, and its thickness should not exceed 20mm.

图 5.7 所示的两种千斤顶的负载框架由三部分组成：负载稳定装置、反作用装置和观察装置。负载稳定装置包括压板、立柱、加载千斤顶和稳定器；反作用装置包括锚系统或桩系统等；观察装置包括千分表和固定支架。载荷测试点通常布置在取样的技术钻孔附近。地质构造简单时，距离不应超过 10m，其他情况下，不应超过 5m，但不应小于 2m。必须注意保持试验土层的原始结构和自然湿度。建议用粗砂或中砂层平整待测表面，其厚度不应超过 20mm。

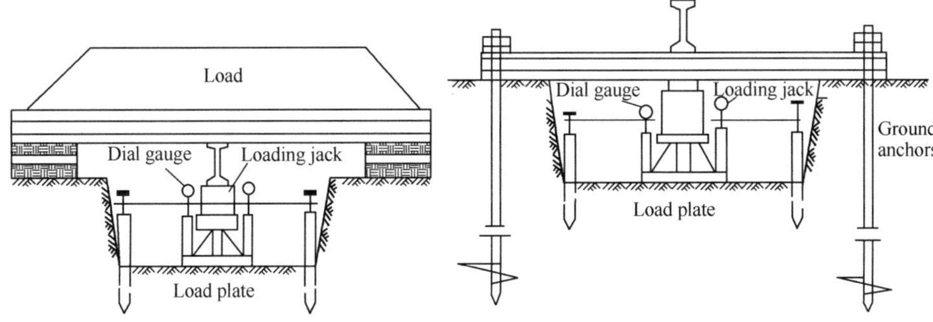

Figure 5.7 Shallow plate load test equipment

The total load applied in the load test should be as close as possible to the ultimate load p_u of the subsoil. The loading level should not be less than 8 levels, and the maximum loading capacity should not be less than twice the design requirements. The first level load (including equipment weight) should be close to the weight of the soil removed by excavating the shallow test pit, and the corresponding settlement of the bearing plate should not be considered; Afterwards, for each level of load increment, 10～25kPa can be used for softer soil, and 50～100kPa can be used for harder and denser soil. The final level of load is the key to determining the bearing capacity, and it should be subdivi-

试验中施加的总荷载应尽可能接近地基的极限荷载 p_u。加载水平不应低于 8 级，最大加载能力不应低于设计要求的两倍。一级荷载（包括设备重量）应接近开挖浅试坑所移除土的重量，不考虑承重板的相应沉降；之后，对于每一级荷载增量，对于较软的土 10～25kPa，较硬和较密的土 50～100kPa。最终荷载水平是确定承载力的关键，应将其细分为两个荷载水平，以提高结果的准确性。最大荷载量不应

ded into two levels of loading to improve the accuracy of the results. The maximum loading amount should not be less than twice the design load value.

Observation criteria for load testing: ① After each level of loading, the settlement amount should be read every 0.5 hours at intervals of 10 minutes, 10 minutes, 10 minutes, 15 minutes, 15 minutes, and beyond. If the settlement rate is less than 0.1mm/h for 2 consecutive hours, it is considered stable and the next level of load can be added. ② When one of the following situations occurs, the loading can be terminated: the soil around the pressure plate has obvious lateral extrusion (sandy soil) or cracks occur (cohesive soil and silt); The settlement s increases sharply, and the load settlement (p-s) curve shows a steep drop section; Under a certain level of load, the settlement rate cannot reach the stability standard within 24 hours; $s/b \geqslant 0.06$ (where b is the width or diameter of the pressure plate).

When one of the first three conditions for terminating the load is met, the corresponding previous level load is determined as the ultimate load.

Based on the observed values of various levels of loads and their corresponding relative stable settlements, the relationship curve between load p and stable settlement s, p-s curve, can be drawn using an appropriate scale. If necessary, the relationship curve between settlement and time under various levels of loads, i.e. s-t curve, can also be drawn. Figure 5.8 shows the typical p-s curve of subsoil. The beginning of the curve is often close to the straight line, and the load p_1 or p_{cr} corresponding to the end point 1 of the straight-line segment is called the proportional limit load of the subsoil. Generally, the allowable bearing capacity or characteristic value of the foundation bearing capacity is taken to be close to this proportional limit load (see Section 9.4 for details).

小于设计荷载值的两倍。

载荷试验的观测标准：①每级荷载后，应每隔 0.5h 读取一次沉降量，间隔时间为 10min、10min、10min、15min、15min 及以上。如果连续 2h 沉降速率小于 0.1mm/h，则认为稳定，可以增加下一级荷载。②当出现以下情况之一时，可以终止加载：压板周围的土有明显的侧向挤压（砂土）或出现裂缝（黏性土和粉土）；沉降 s 急剧增加，荷载-沉降（p-s）曲线呈陡降段；在一定荷载水平下，沉降率在 24h 内无法达到稳定标准；$s/b \geqslant 0.06$（其中 b 为压板的宽度或直径）。

当满足终止负载的前三个条件之一时，相应的前一级荷载被确定为极限荷载。

根据各级荷载的观测值及其相应的相对稳定沉降，可以使用适当的比例绘制荷载 p 和稳定沉降 s 之间的关系曲线，即 p-s 曲线。如有必要，还可以绘制不同荷载水平下沉降与时间的关系曲线，即 s-t 曲线。图 5.8 为土的典型 p-s 曲线。曲线的起点通常接近直线，与直线段终点 1 对应的荷载 p_1 或 p_{cr} 称为土的比例极限荷载。通常，允许承载力或基础承载力特征值被视为接近该比例极限荷载（详见第 9.4 节）。

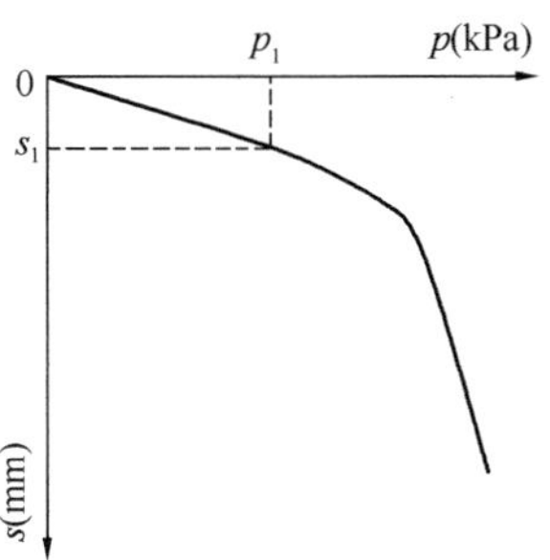

Figure 5.8 Typical p-s curve

So, the deformation of the subsoil is in the stage of linear deformation. Therefore, the elastic mechanics equation for surface settlement of the subsoil (Eq. 5.2) $s=\omega(1-\mu^2)bp/E_0$, p is the uniformly distributed load on the subsoil surface (kPa), and ω is the settlement influence coefficient. The deformation modulus of the subsoil should be calculated by taking $\omega_r=0.886$ (square pressure plate) or 0.785 (circular pressure plate) for rigid pressure plates. The calculation equation is as follows:

因此，地基土的变形处于线性变形阶段。所以，地基表面沉降的弹性力学方程［式（5.2）］可写为 $s=\omega(1-\mu^2)bp/E_0$，p 是地基表面均匀分布的荷载（kPa），ω 是沉降影响系数。地基的变形模量应按 $\omega_r=0.886$（方形压力板）或 0.785（圆形压力板）计算，对于刚性压力板。计算公式见式（5.15）～式（5.17）。

$$E_0=0.886(1-\mu^2)bp_1/s_1 \tag{5.15}$$

$$E_0=0.785(1-\mu^2)dp_1/s_1 \tag{5.16}$$

or

$$E_0=(1-\mu^2)P/s_1d \tag{5.17}$$

where E_0 is the deformation modulus of soil; μ is the Poisson's ratio of soil; b is the side length of the pressure plate (m); d is the diameter of the pressure plate (m); P is the load of the pressure plate (kN); p_1 is the determined proportional limit load (kPa), $p_1=P/b^2$ or $p_1=4P/\pi d^2$; s_1 is the settlement (mm) corresponding to the proportional limit load p_1. Sometimes, the p-s curve does not have a starting straight-line segment, so s_1/b or $s_1/d=0.010\sim0.015$ (low value for low compressibility soil and high value for high compressibility soil) and its corresponding load p_1 can be substituted into the equation.

式中，E_0 是土的变形模量；μ 是土的泊松比；b 为压板边长（m）；d 为压板直径（m）；P 为压板荷载（kN）；p_1 是确定的比例极限载荷（kPa），$p_1=P/b^2$ 或 $p_1=4P/\pi d^2$；s_1 是与比例极限载荷 p_1 对应的沉降（mm）。有时，p-s 曲线没有起始直线段，因此可将 s_1/b 或 $s_1/d=0.010\sim0.015$（低压缩性土为低值，高压缩性土为高值）及其相应的荷载 p_1 代入公式。

In addition, the deformation modulus of the foundation soil can also be obtained from deep plate loading

此外，地基土的变形模量也可通过深层平板载荷试验和旁压

tests and pressuremeter tests. The experimental process and method can refer to the " Code for Investigation of Geotechnical Engineering" GB 50021—2001 (2009 edition) and the " Code for Design of Building Foundation" GB 50007—2011.

试验获得。试验过程和方法可参考《岩土工程勘察规范》GB 50021—2001（2009 年版）和《建筑地基基础设计规范》GB 50007—2011。

5.4.2 The Relationship Between Deformation Modulus and Compression Modulus

5.4.2 变形模量与压缩模量的关系

As mentioned earlier, the deformation modulus E_0 of soil is the ratio of stress to strain of the soil under unconfined conditions, while the compression modulus E_s of soil is the ratio of stress to strain of the soil under lateral conditions. E_0 and E_s can theoretically be converted to each other.

如前所述，土的变形模量 E_0 是无侧限条件下土的应力应变比，而土的压缩模量 E_s 是侧限条件下土应力与应变的比。E_0 和 E_s 理论上可以相互转换。

Take a micro unit from the sample the consolidation test. Under the pressure in the z-axis direction, the vertical normal stress in the sample is σ_z, and since the stress conditions of the sample belong to the axisymmetric problem, the corresponding horizontal normal stress is $\sigma_x=\sigma_y$, as shown in Eq. (5.18)

在固结试验中，从试样中取出一个微单元。在 z 轴方向的压力下，试样中的垂直法向应力为 σ_z，由于试样的应力条件属于轴对称问题，相应的水平法向应力为 $\sigma_x = \sigma_y$，如式（5.18）所示，

$$\sigma_x = \sigma_y = K_0\sigma_z \tag{5.18}$$

where K_0 is the lateral pressure coefficient of the soil.

式中，K_0 是土的侧向压力系数。

According to the generalized Hooke's law,

根据广义胡克定律，可以得出式（5.19）和式（5.20）。

$$\varepsilon_x = \frac{\sigma_x}{E_0} - \mu\frac{\sigma_y}{E_0} - \mu\frac{\sigma_z}{E_0} \tag{5.19}$$

$$\varepsilon_z = \frac{\sigma_z}{E_0} - \mu\frac{\sigma_y}{E_0} - \mu\frac{\sigma_x}{E_0} \tag{5.20}$$

Due to the sample not allowing lateral expansion, $\varepsilon_x=\varepsilon_y=0$. Therefore,

由于试样不允许横向膨胀，$\varepsilon_x=\varepsilon_y=0$，因此，有式（5.21）。

$$\varepsilon_x = \frac{\sigma_x}{E_0} - \mu\frac{\sigma_y}{E_0} - \mu\frac{\sigma_z}{E_0} = 0 \tag{5.21}$$

Substituting Eq. (5.18) into Eq. (5.20) yields

将式（5.18）代入式（5.20）得出 K_0 和 μ 的表达式。

$$K_0 = \mu/(1-\mu)$$

or

$$\mu = K_0/(1+K_0)$$

According to the lateral confinement conditions, $\varepsilon_z=\sigma_z/E_s$, Therefore,

根据横向约束条件，$\varepsilon_z = \sigma_z/E_s$，

$$E_0 = \beta E_s \tag{5.22}$$

where

$$\beta = 1 - 2\mu K_0 = (1+\mu)(1-2\mu)/(1-\mu) \quad (5.23)$$

It must be pointed out that the above equation is only a theoretical relationship between E_0 and E_s. In fact, due to some factors that cannot be considered when measuring E_0 in on-site load tests and E_s in indoor compression tests, the above equation cannot accurately reflect the actual relationship between E_0 and E_s.

因此，有式（5.22），式中，β 的表达式为式（5.23）。必须指出，上述方程只是 E_0 和 E_s之间的理论关系。事实上，由于现场载荷试验测量 E_0 和室内压缩试验测量 E_s时无法考虑一些因素，上述公式无法准确反映 E_0 和 E_s 之间的实际关系。

Problems

5.1 Refer to Figure 5.9. Determine the vertical stress increase, $\Delta\sigma_z$ at point A with the following values: q_1=90kN/m; q_2=325kN/m; x_1=4m; x_2=2.5m; z=3m.

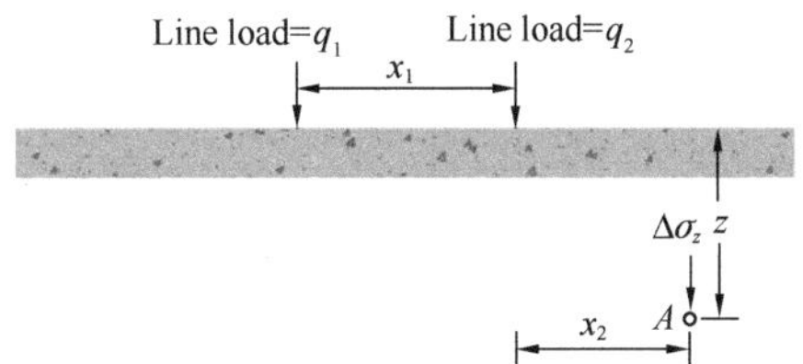

Figure 5.9 Figure of problem 5.1

5.2 Refer to Figure 5.10. A strip load of q=43kN/m^2 is applied over a width, B=11m. Determine the increase in vertical stress at point A located z=4.6m below the surface. Given: x=8.2m.

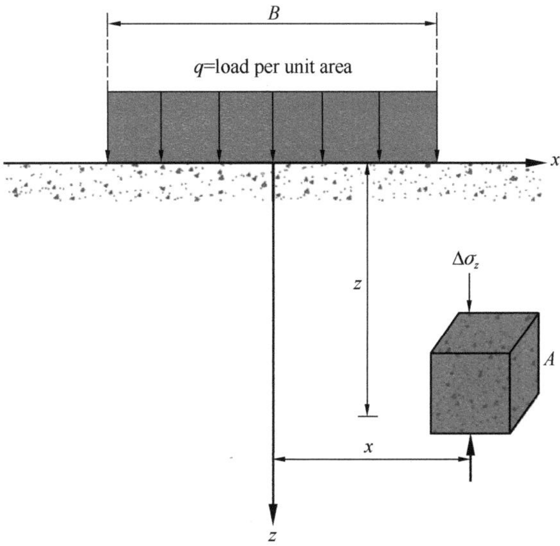

Figure 5.10 Figure of problem 5.2

5.3 An earth embankment is shown in Figure 5.11. Determine the stress increase at point A due to the embankment load.

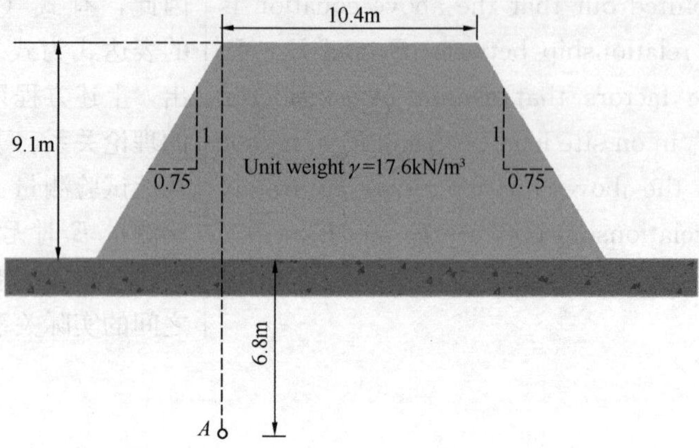

Figure 5.11 Figure of problem 5.3

5.4 Refer to Figure 5.12. A flexible rectangular area is subjected to a uniformly distributed load of $q=225\text{kN/m}^2$. Determine the increase in vertical stress, $\Delta\sigma_z$ at a depth of $z=3\text{m}$ under points A, B, and C.

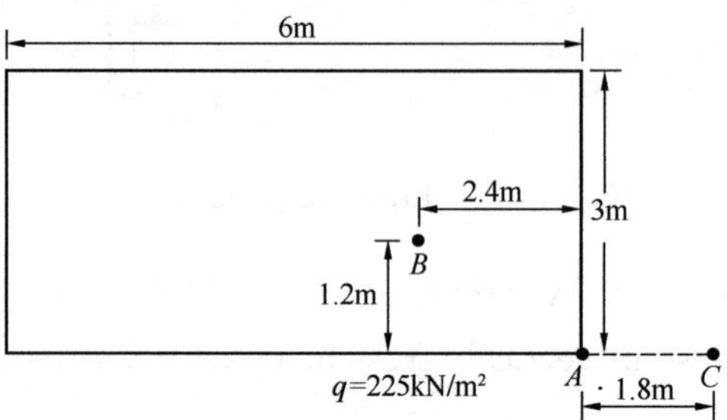

Figure 5.12 Figure of problem 5.4

Chapter 6 Deformation of Subsoil

6.1 Introduction

According to Chapter 5, additional stress in the subsoil can cause soil particles to compress and displace each other, resulting in a macroscopic reduction or expansion of soil volume, known as deformation, including settlement and displacement. This chapter will focus on the settlement of the ground surface and will cover procedure for calculating elastic settlement, layered summation method for calculating final settlement, Terzaghi's one-dimensional consolidation theory, primary consolidation settlement and secondary consolidation settlement.

6.2 Elastic Settlement

In Chapter 4, the relationships for determining the additional stress (which causes elastic settlement) due to the application of line load, strip load, circular load, and rectangular load were based on the following assumptions:

(1) The load is applied at the ground surface.

(2) The loaded area is flexible.

(3) The soil medium is homogeneous, elastic, isotropic, and extends to a great depth.

Therefore, the solution for the vertical displacement $w(x, y, z)$ generated at any point $M(x, y, z)$ in an elastic half space when a vertically concentrated force P acts on the surface of the half space can be obtained from the Boussinesq fundamental solution. If $z=0$, the vertical displacement $w(x, y, z)$ of any point on the surface of the half space obtained can be regarded as the settlement s_e of any point on the ground surface (Figure 6.1).

第6章 地基变形

6.1 概述

根据第5章，地基土中的附加应力会导致土颗粒相互压缩和位移，从而在宏观上导致土体积的减小或膨胀，称为变形，包括沉降和位移。本章将重点介绍地表沉降，并将涵盖计算弹性沉降的步骤、计算最终沉降的分层总和法、太沙基的一维固结理论、主固结沉降和次固结沉降。

6.2 弹性沉降

在第4章中，基于以下假设确定由于施加线荷载、条形荷载、圆形荷载和矩形荷载而产生的附加应力（导致弹性沉降）的关系：

（1）荷载施加在地面上；

（2）荷载施加区域是柔性的；

（3）土介质是均匀的、弹性的、各向同性的，并且延伸到很深的地方。

因此，根据布辛奈斯克基本解，当竖向集中力 P 作用于弹性半空间的表面时，在弹性半空间中的任何点 $M(x, y, z)$ 处产生的竖向位移 $w(x, y, z)$ 的解可以从布辛奈斯克基本解中获得。如果 $z=0$，则所获得的半空间表面上任何点的竖向位移 $w(x, y, z)$ 都可以视为地面上任一点的沉降 s_e（图6.1）。

$$s_e = w(x,y,0) = \frac{P(1-\mu^2)}{\pi E_0 r} \tag{6.1}$$

where s_e is the elastic settlement of ground surface at any point under the action of vertical concentrated force P; r is the distance from any point on the foundation surface to the point of vertical concentrated force application, $r = (x^2+y^2)^{0.5}$; E_0 is the deformation modulus of soil; μ is Poisson's ratio of soil.

式中，s_e是竖向集中力 P 作用下任意点地表的弹性沉降；r 为基础表面上任意一点到垂直集中力施加点的距离 $r = (x^2 + y^2)^{0.5}$；E_0 为土体变形模量；μ 是土的泊松比。

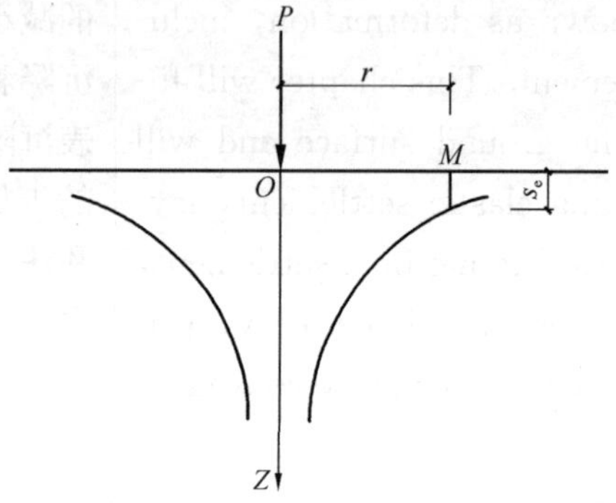

Figure 6.1　Curve of the settlement of ground surface under vertical concentrated force

As shown in Figure 6.2, assuming the distributed load at any point $N(\xi, \eta)$ on the load surface A is $p(\xi, \eta)$, the distributed load on the micro area $d\xi d\eta$ at that point can be replaced by the concentrated force $P = p(\xi, \eta)\, d\xi d\eta$. Therefore, the settlement $s_e(x, y)$ at point $M(x, y)$, which is located at a distance of $r = [(x-\xi)^2 + (y-\eta)^2]^{0.5}$ from the point of vertical concentrated force, can be obtained by integrating according to Eq. (6.1),

如图 6.2 所示，假设载荷面 A 上任何点 $N(\xi, \eta)$ 处的分布荷载为 $p(\xi, \eta)$，则该点微面积 $d\xi d\eta$ 上的分布荷载可以用集中力 $P = p(\xi, \eta)\, d\xi d\eta$ 表示。因此，通过根据方程（6.1）进行积分，可以获得位于距离竖向集中力点 $r = [(x-\xi)^2 + (y-\eta)^2]^{0.5}$ 处的点 $M(x, y)$ 的沉降 $s_e(x, y)$ 计算式（6.2）。

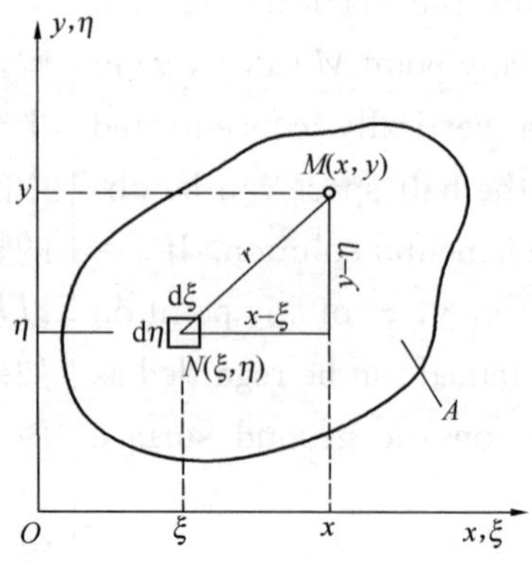

Figure 6.2　Calculation of the settlement of ground surface under distributed load

$$s_e(x,y)=\frac{(1-\mu^2)}{\pi E_0}\iint_A\frac{p(\xi,\eta)\mathrm{d}\xi\mathrm{d}\eta}{\sqrt{(x-\xi)^2+(y-\eta)^2}} \tag{6.2}$$

For uniformly distributed rectangular loads, $p(\xi,\eta)=p$. The settlement generated at the corner point of the rectangle is expressed as the result of integrating according to Eq. (6.2):

对于均匀分布的矩形荷载，$p(\xi,\eta)=p$。矩形角点产生的沉降是根据式（6.2）的积分结果表示的，见式（6.3）。

$$s_e=\omega_c\frac{1-\mu^2}{E_0}bp \tag{6.3}$$

where ω_c is the settlement influence coefficient at corner point. It is a function of the rectangular length-width ratio $m=l/b$.

式中，ω_c是角点处的沉降影响系数。它是矩形长宽比 $m=l/b$ 的函数，它可表达为式（6.4）。

$$\omega_c=\frac{1}{\pi}\left[m\ln\frac{1+\sqrt{m^2+1}}{m}+\ln(m+\sqrt{m^2+1})\right] \tag{6.4}$$

It is easy to obtain the settlement of any point on the foundation surface under uniformly distributed rectangular loads using the corner method (similar to the method shown in Figure 4.13 in Section 4.4.2).

For the settlement of the center point of the rectangle, it is equal to the sum of the settlement of dividing it into four identical small rectangular corner points, that is

使用角点法（类似于第4.4.2节图4.13所示的方法）很容易获得均匀分布的矩形荷载下基础表面上任意点的沉降。

对于矩形中心点的沉降，它等于将其划分为四个相同的小矩形角点的沉降之和，如式（6.5）所示。

$$s_e=4\omega_c\frac{1-\mu^2}{E_0}(b/2)p=2\omega_c\frac{1-\mu^2}{E_0}bp \tag{6.5}$$

By comparing Eqs. (6.3) and (6.5), it can be known that the settlement of the center point of a rectangular load is twice the settlement of the corner points. If $\omega_0=2\omega_c$, it is called the influence coefficient of the center point settlement and yields

通过比较式（6.3）和式（6.5），可以知道矩形荷载中心点的沉降是角点沉降的两倍。令$\omega_0=2\omega_c$，称为中心点的沉降影响系数，并得出式（6.6）。

$$s_e=\omega_0\frac{1-\mu^2}{E_0}bp \tag{6.6}$$

6.3 Final Settlement of sub soil

6.3 地基最终沉降

The final settlement of the subsoil under foundation refers to the settlement of the foundation bottom surface after the subsoil is deformed and stabilized under the

地基的最终沉降是指地基在建筑物等其他荷载作用下变形稳定后，基础底面的沉降。计算基

action of other loads such as buildings. There are usually two methods for calculating the final settlement of a foundation bottom surface: the layered sum method and the standard recommendation method.

础底面的最终沉降通常有两种方法：分层总和法和规范推荐法。

6.3.1 Layered Sum Method

The basic principle of the layered sum method is to divide the depth range of the foundation compression layer into several layers, calculate the compression amount of each layer, and then calculate the sum. The depth of compressive layer refers to the depth at which the deformation needs to be calculated from the bottom surface of the foundation. Below this depth, the deformation value of the soil layer is so small that it can be ignored, also known as the depth of subsoil deformation calculation.

To derive the theoretical equation for unidirectional compression using the layered summation method, the assumptions are made as following:

(1) The contact pressure under a foundation is linearly distributed.

(2) Additional stress is calculated according to elastic theory.

(3) Compression only occurs vertically.

(4) The calculation is only for consolidation settlement.

As shown in Figure 6.3, apply a continuous and uniform load p_c on a soil layer with a thickness of H_1, generating vertical additional stress in the soil.

6.3.1 分层总和法

分层总和法的基本原理是将基础压缩层在计算深度范围内划分为几层，计算每层的压缩量，然后计算总和。压缩层深度是指从基础底面开始需要计算变形的深度。在此深度以下，土层的变形值很小，可以忽略不计，也称为地基变形计算深度。

为了使用分层总和法推导单向压缩的理论方程，假设如下：

（1）基础下的接触压力呈线性分布。

（2）根据弹性理论计算附加应力。

（3）仅竖向可发生压缩。

（4）该计算仅适用于固结沉降。

如图 6.3 所示，在厚度为 H_1 的土层上施加连续均匀的荷载 p_c，在土中产生竖向附加应力。

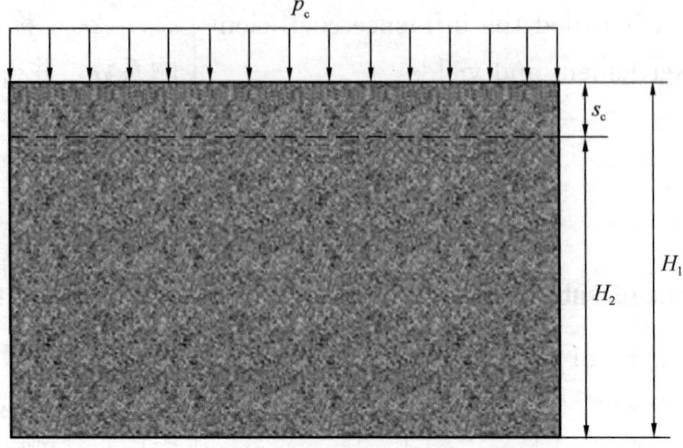

Figure 6.3 Analysis of settlement of single-layer soil

Assuming that the stress in the soil increases from the original self-weight stress p_1 to the sum of self-weight stress and additional stress p_2, the soil is compressed, the height of the soil layer decreases from H_1 to H_2 and the void ratio decreases from e_1 before compression to e_2 after compression stability. According to Eq. (5.5), yields

假设土中的应力从原始自重应力 p_1 增大到自重应力和附加应力之和 p_2，土体被压缩，土层高度从 H_1 减小到 H_2，孔隙比从压缩前的 e_1 减小到压缩稳定后的 e_2。根据式（5.5），有式（6.7），

$$\frac{H_2}{1+e_2}=\frac{H_1}{1+e_1} \tag{6.7}$$

The compression amount of this soil layer is

该土层的压缩量为式（6.8）。

$$s_c=H_1-H_2=\frac{e_1-e_2}{1+e_1}H_1 \tag{6.8}$$

According to the relationship between compression coefficient and compression modulus, Eq. (6.8) can be changed to

根据压缩系数和压缩模量之间的关系，式（6.8）可改写为式（6.9），

$$s_c=\frac{a\Delta p}{1+e_1}H_1=\frac{\Delta p}{E_s}H_1 \tag{6.9}$$

where Δp is average additional stress within the thickness of the soil layer, $\Delta p=p_1-p_2$; a and E_s are respectively compression coefficient and compression modulus of the soil layer.

式中，Δp 是土层厚度内的平均附加应力，$\Delta p=p_1-p_2$；a 和 E_s分别为土层的压缩系数和压缩模量。

If the compressible soil layer is divided into n layers (Figure 6.4), the settlement of the compressed layer is

如果可压缩土层分为 n 层（图 6.4），则压缩层的沉降可表示为式（6.10），

$$s_c=\sum_{i=1}^{n}\Delta s_{ci}=\sum_{i=1}^{n}\frac{e_{1i}-e_{2i}}{1+e_{1i}}H_i=\sum_{i=1}^{n}\frac{a_i(p_{2i}-p_{1i})}{1+e_{1i}}H_i=\sum_{i=1}^{n}\frac{\Delta p_i}{E_{si}}H_i=\sum_{i=1}^{n}\varepsilon_i H_i \tag{6.10}$$

where Δs_{ci} is the compression deformation of the i-th layer of soil; $\varepsilon_i=\Delta p_i/E_{si}$ is the compressive strain of the i-th layer of soil; H_i is the thickness of the i-th layer of soil; e_{1i} is the void ratio corresponding to the average self-weight stress p_{1i} of the i-th layer of soil on the e-p curve; e_{2i} is the void ratio corresponding to the sum of the average self-weight stress and the average additional stress p_{2i} of the i-th layer of soil on the e-p curve.

式中，Δs_{ci}是第 i 层土的压缩变形；$\varepsilon_i=\Delta p_i/E_{si}$ 是第 i 层土的压缩应变；H_i是第 i 层土的厚度；e_{1i}是 e-p 曲线上第 i 层土的平均自重应力 p_{1i}对应的孔隙比；e_{2i}是对应于 e-p 曲线上第 i 层土的平均自重应力和平均附加应力之和 p_{2i}的孔隙比。

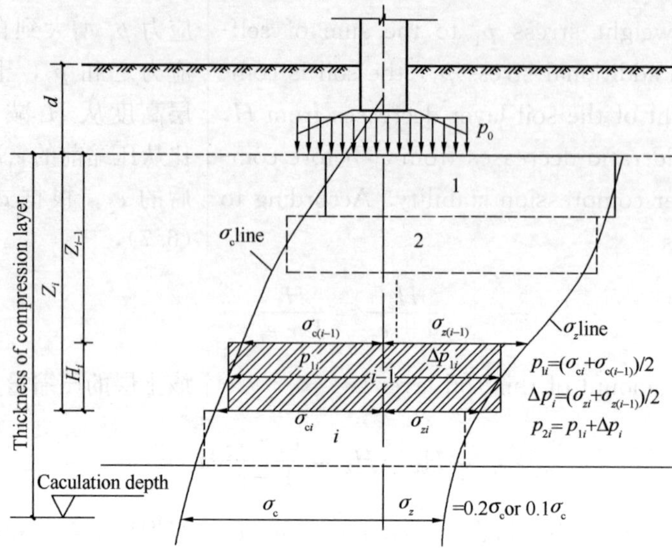

Figure 6.4 Calculation principle of layered sum method

When calculating, the depth of compression layer must be determined first, and layering must be carried out within this depth range. Then, the average self-weight stress and average additional stress of each point on the top and bottom surfaces of the layering under the central axis of the foundation must be calculated. The lower limit of the depth of the compression layer is taken at the point where the additional stress is equal to 20% of the self-weight stress, that is, at $\sigma_z=0.2\sigma_c$; If there is a highly compressible soil layer below this depth, the calculation should continue downwards until $\sigma_z=0.1\sigma_c$. The layer thickness within the depth range of the compression layer can be taken as about 0.4b (b is the width of the short side of the foundation), and the layers of the layered soil and the groundwater level are both natural layer surfaces.

计算时，必须首先确定压缩层的厚度，并在该深度范围内进行分层。然后，须计算基础中心下分层顶面和底面上每个点的平均自重应力和平均附加应力。压缩层深度的下限取附加应力等于自重应力20%的点，即$\sigma_z=0.2\sigma_c$；如果在此深度以下存在高压缩性的土层，则应继续向下计算，直到$\sigma_z=0.1\sigma_c$。压缩层深度范围内的层厚可取约0.4b(b为基础短边宽度)，分层土层和地下水位均为天然层面。

6.3.2 Revised Layered Sum Method

The concept of average additional stress coefficient is proposed in the " Code for Design of Building Foundation" GB 50007—2011, which is defined as the ratio of the area of additional stress (distribution map) A to the product of the additional pressure on the foundation bottom and the depth of the subsoil under the foundation,

6.3.2 修正的分层总和法

《建筑地基基础设计规范》GB 50007—2011 中提出了平均附加应力系数的概念，其定义为基底某点下至地基任意深度范围内的附加应力面积（分布图）A与基础底面附加压力和地基深度

$p_0 z$, from a certain point below the foundation to any depth range of the subsoil (Figure 6.5).

的乘积 $p_0 z$ 的比值（图 6.5）。

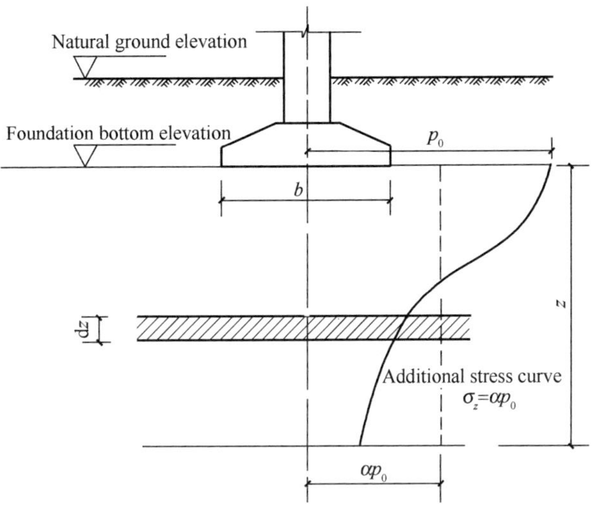

Figure 6.5 Definition of average additional stress coefficient

Therefore, the average additional stress coefficient can be expressed as

因此，平均附加应力系数可以表示为式（6.11）。

$$\bar{\alpha} = A/p_0 z \tag{6.11}$$

Assuming that the subsoil under foundation is homogeneous and the compression modulus E_s under lateral confinement conditions does not vary with depth, the compression deformation s'_c from the foundation bottom to the depth z below it can be expressed as:

假设地基土是均匀的，侧限条件下的压缩模量 E_s 不随深度变化，则从基底至其下方深度 z 的压缩变形 s'_c 可以表示为式（6.12），

$$s'_c = \int_0^z \varepsilon \mathrm{d}z = \frac{1}{E_s}\int_0^z \sigma_z \mathrm{d}z = \frac{A}{E_s} \tag{6.12}$$

where ε is the compression strain of soil, $\varepsilon = \sigma_z/E_s$; σ_z is vertical additional stress in soil; A is the additional stress area from a certain point on the bottom of the foundation to any depth z below it.

式中，ε 是土的压缩应变，$\varepsilon = \sigma_z/E_s$；$\sigma_z$ 是土中的竖向附加应力；A 是基底某点到其下方任意深度 z 范围内的附加应力面积。A 可表达为式（6.13）。

$$A = \int_0^z \sigma_z \mathrm{d}z = p_0 \int_0^z \alpha \mathrm{d}z \tag{6.13}$$

It can be derived from Eqs. (6.11) and (6.12) that

从式（6.11）和式（6.12）可以得出式（6.14）。

$$s'_c = p_0 z\bar{\alpha}/E_s \tag{6.14}$$

where $\bar{\alpha}$ is the average vertical additional stress coefficient within the depth range of z; $p_0 z\bar{\alpha}$ represents the area enclosed by the additional stress curve within the

式中，$\bar{\alpha}$ 是 z 深度范围内的平均竖向附加应力系数；$p_0 z\bar{\alpha}$ 表示 z 深度范围内附加应力曲线所包围

depth range of z.

Figure 6.6 shows the calculation principle of revised layered sum method proposed in the "Code for Design of Building Foundation" GB 50007—2011. Therefore, according to Eq. (6.14), the vertical compression deformation equation for the i-th layer in a layered subsoil under foundation can be obtained that

的面积。

图 6.6 展示了《建筑地基基础设计规范》GB 50007—2011 中提出的修正分层总和法的计算原理。因此，根据式（6.14），可以得出成层地基中第 i 分层的竖向压缩变形公式（6.15），

$$\Delta s'_i = s'_i - s'_{i-1} = \frac{A_i - A_{i-1}}{E_{si}} = \frac{\Delta A_i}{E_{si}} = \frac{p_0}{E_{si}}(z_i\bar{\alpha}_i - z_{i-1}\bar{\alpha}_{i-1}) \tag{6.15}$$

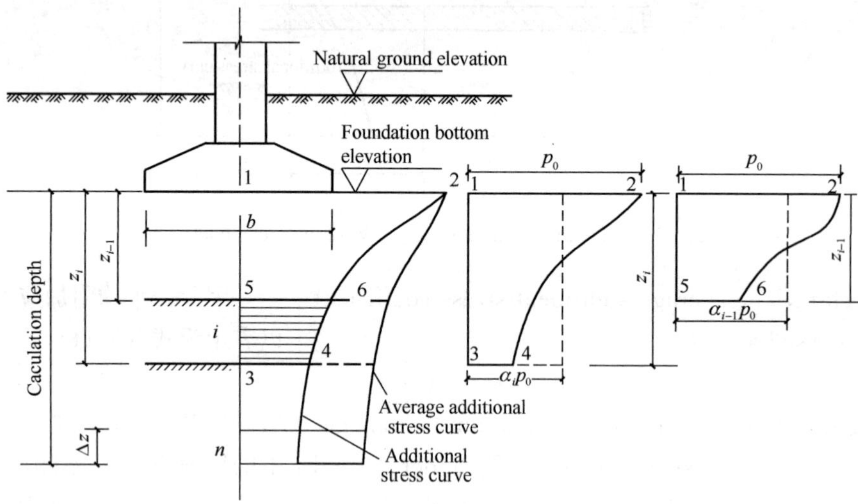

Figure 6.6　Calculation principle of revised layered sum method

where s'_i and s'_{i-1} are respectively deformation within the depth range of z_i and z_{i-1}; $\bar{\alpha}_i$ and $\bar{\alpha}_{i-1}$ are respectively average vertical additional stress coefficient within the depth range of z_i and z_{i-1}, they can be obtained by checking the additional stress coefficient table in the "Code for Design of Building Foundation" GB 50007—2011 based on the values of l/b and z/b; $p_0 z_i \bar{\alpha}_i$ is equivalent value of the area A_i enclosed by the additional stress curve (Area 1234 in the Figure 6.6) within the depth range of z_i; $p_0 z_{i-1} \bar{\alpha}_{i-1}$ is equivalent value of the area A_{i-1} enclosed by the additional stress curve (Area 1256 in the Figure 6.6) within the depth range of z_{i-1}; ΔA_i is the area enclosed by the vertical additional stress curve of the i-th layer (Area 5634 in the Figure 6.6).

式中，s'_i 和 s'_{i-1} 分别是 z_i 和 z_{i-1} 深度范围内的变形；$\bar{\alpha}_i$ 和 $\bar{\alpha}_{i-1}$ 分别为 z_i 和 z_{i-1} 深度范围内的平均竖向附加应力系数，可根据 l/b 和 z/b 的值，通过查阅《建筑地基基础设计规范》GB 50007—2011 中的附加应力系数值表获得；$p_0 z_i \bar{\alpha}_i$ 是 z_i 深度范围内附加应力曲线所包围的面积 A_i（图 6.6 中的面积 1234）的等效值；$p_0 z_{i-1} \bar{\alpha}_{i-1}$ 是 z_{i-1} 深度范围内附加应力曲线（图 6.6 中的面积 1256）包围的面积 A_{i-1} 的等效值；ΔA_i 是第 i 层垂直附加应力曲线所包围的面积（图 6.6 中的 5634 区域）。

Adding up the deformations of each layer yields the total settlement of the compressed layer, that is,

将每层的变形加起来即得压缩层的总沉降公式（6.16）。

$$s'_c = \sum_{i=1}^{n} \Delta s'_i = \sum_{i=1}^{n} \frac{p_0}{E_{si}} (z_i \bar{\alpha}_i - z_{i-1} \bar{\alpha}_{i-1}) \tag{6.16}$$

The determination of the depth of the compression layer, also known as the deformation calculation depth, is based on taking the calculated thickness Δz specified in Table 6.1 upwards from that depth (Figure 6.6). The resulting deformation $\Delta' s_n$ should meet the requirements of the following equation:

确定压缩层的深度，也称为变形计算深度，是从该深度向上取表 6.1 中规定的计算厚度 Δz（图 6.6）。由此产生的变形 $\Delta' s_n$ 应满足式（6.17）的要求。

$$\Delta s'_n \leqslant 0.025 \sum_{i=1}^{n} \Delta s'_i \tag{6.17}$$

The value of calculation depth Δz **Table 6.1**

b (m)	≤2	2<b≤4	4<b≤8	>8
Δz (m)	0.3	0.6	0.8	1.0

When there is no adjacent load influence and the foundation width is within the range of 1～30m, the calculation depth of the subsoil deformation at the midpoint of the foundation can also be calculated according to the following simplified equation:

当无相邻荷载影响且基础宽度在 1～30m 范围内时，基础中心点地基土变形的计算深度也可以根据简化公式（6.18）计算。

$$z_n = b(2.5 - 0.4 \ln b) \tag{6.18}$$

where b is width of the foundation.

式中，b 是基础的宽度。

When there is a bedrock layer within the depth range of deformation calculation, z_n can be taken up to the surface of the bedrock. When there is a thick hard viscous soil layer with a void ratio less than 0.5 and a compression modulus greater than 50MPa, or when there is a thick dense sand and gravel layer with a compression modulus greater than 80MPa, z_n can be taken up to the surface of the soil layer.

当变形计算深度范围内有基岩层时，z_n 可算至基岩表面。当存在孔隙比小于 0.5 且压缩模量大于 50MPa 的厚硬黏性土层时，或者存在压缩模量大于 80MPa 的厚密砂砾石层时，z_n 可算至该土层表面。

In order to improve the accuracy of calculations, the foundation design specification proposes an empirical coefficient for settlement calculation, which is defined as

为了提高计算的准确性，地基基础设计规范提出了一个用于计算沉降的经验系数，其定义为式（6.19）。

$$\phi_s = s_\infty / s'_c \tag{6.19}$$

where s_∞ is final settlement prediction based on measured data; ϕ_s is an empirical coefficient for settlement calculation, it can be obtained from Table 6.2.

式中，s_∞ 是基于实测数据推算的最终沉降量；ϕ_s 是沉降计算的经验系数，可从表 6.2 中获得。

Empirical coefficient for settlement calculation **Table 6.2**

p_0 (MPa)	E_s (MPa)				
	2.5	4.0	7.0	15.0	20.0
$p_0 \geqslant f_{ak}$	1.4	1.3	1.0	0.4	0.2
$p_0 \leqslant 0.75 f_{ak}$	1.1	1.0	0.7	0.4	0.2

where f_{ak} is characteristic value of bearing capacity of bearing layer; $\bar{E}_s$ is equivalent value of compression modulus within the depth range of settlement calculation,

式中，f_{ak} 为地基承载力特征值；$\bar{E}_s$ 是沉降计算深度范围内的压缩模量的当量值，可表达为式（6.20），

$$\bar{E}_s = \sum \Delta A_i \Big/ \sum \Delta A_i / E_{si} \tag{6.20}$$

Therefore, the revised final settlement equation for the subsoil is as follows:

因此，修正的地基最终沉降量为式（6.21）。

$$s_c = \phi_s s'_c = \phi_s \sum_{i=1}^{n} \frac{p_0}{E_{si}} (z_i \bar{\alpha}_i - z_{i-1} \bar{\alpha}_{i-1}) \tag{6.21}$$

where s_c is final settlement of the subsoil; s'_c is the deformation of subsoil calculated by the modified layered sum method; n is the number of soil layers divided within the depth range for deformation calculation. The layer and groundwater level are natural layering surfaces, and in order to improve the accuracy of E_{si} values, the layer thickness should not exceed 2m; E_{si} is compressive modulus of i-th layer soil under the foundation bottom (MPa); z_i and z_{i-1} are respectively distance from the foundation bottom to the i-th layer of soil and the i-1-th layer of soil bottom (m) .

式中，s_c 是地基的最终沉降；s'_c 是用修正的分层总和法计算的地基变形；n 是变形计算深度范围内划分的土层数。层面和地下水位面是自然分层面，为了提高 E_{si} 值的准确性，层厚不应超过 2m；E_{si} 为基底下第 i 层土的压缩模量（MPa）；z_i 和 z_{i-1} 分别是从基底到第 i 层土和第 $i-1$ 层土底面的距离（m）。

Example 6.1

The distribution of soil layers, soil parameters, and additional stress curve of a building's are shown in Figure 6.7. The additional stress curve is divided into two sections: $0 \leqslant z \leqslant 3\text{m}$, $f(z) = 120 - 10z^2$; $3\text{m} \leqslant z \leqslant 4\text{m}$, additional stress linearly distributed. Calculate final settlement of the building subsoil.

例题 6.1

建筑物的土层分布、土体参数和附加应力曲线如图 6.7 所示。附加应力曲线分为两段：$0 \leqslant z \leqslant 3\text{m}$，$f(z) = 120 - 10z^2$；$3\text{m} \leqslant z \leqslant 4\text{m}$，附加应力呈线性分布。计算建筑地基的最终沉降。

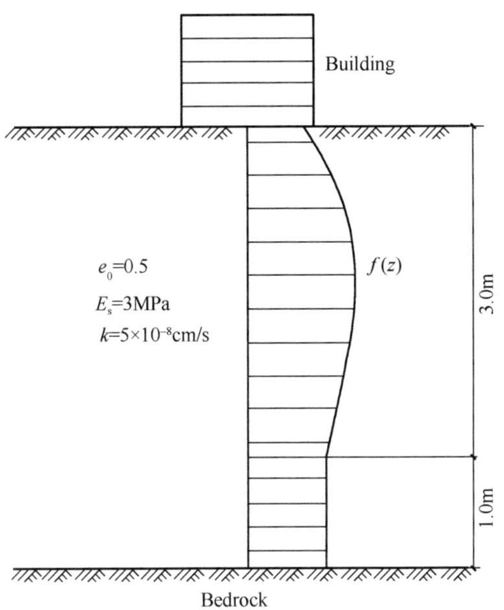

Figure 6.7 Figure of example 6.1

Solution

Divide the compression layer into 2 layers. According to Eq. (6.12), it can be obtained that

解答

将压缩层分为 2 层。根据式(6.12)，可得：

$$s'_{c1} = \frac{1}{E_s}\int_0^z \sigma_z \mathrm{d}z = \frac{1}{3}\int_0^3 (120 - 10z^2)\mathrm{d}z = 90\mathrm{mm}$$

$$s'_{c2} = \frac{1}{E_s}\int_0^z \sigma_z \mathrm{d}z = \frac{1}{3}\int_3^4 30\mathrm{d}z = 10\mathrm{mm}$$

$$s'_c = \sum_{i=1}^{2} s'_{ci} = 90 + 10 = 100\mathrm{mm}$$

6.4 Time Rate of Consolidation

From the above sections, it is known that the deformation of the subsoil increases over time. In engineering design, in addition to knowing the final settlement of the foundation, it is often necessary to know the process of settlement over time, that is, the relationship between settlement and time. In order to arrange the construction sequence reasonably, control the construction speed, and take necessary measures to eliminate adverse effects.

6.4 固结时间速率

从前述章节可以看出，地基土的变形会随着时间的推移而增大。在工程设计中，除了解基础的最终沉降外，通常还需要了解沉降随时间的变化过程，即沉降与时间的关系。合理安排施工顺序，控制施工速度，采取必要措施消除不利影响。

6.4.1 Terzaghi's Theory of one-dimensional Consolidation

In 1925, Terzaghi proposed the first theory to consider the rate of one-dimensional consolidation for satu-

6.4.1 太沙基一维固结理论

1925 年，太沙基提出了第一个考虑饱和黏土一维固结速率

rated clay soils and in 1943, developed an analytical model for determining the degree of consolidation within soil at any time, t. The assumptions made in the theory are:

的理论，并于 1943 年发展了一个解析模型，用于确定地基任意时刻 t 的固结度。该理论基于以下假设：

(1) The soil is homogeneous.

(2) The soil is fully saturated.

(3) The solid particles and water are incompressible.

(4) Compression and flow are one-dimensional (vertical).

(5) Strains are small.

(6) Darcy's law is valid at all hydraulic gradients.

(7) The coefficient of permeability and the coefficient of volume compressibility remain constant throughout the process.

(8) There is a unique relationship, independent of time, between void ratio and effective stress.

(1) 土体是均匀的。

(2) 土体是完全饱和。

(3) 固体颗粒和水是不可压缩的。

(4) 压缩和渗流是一维的（竖向）。

(5) 土体应变很小。

(6) 达西定律适用于所有水力梯度。

(7) 渗透系数和体积压缩系数在整个过程中保持恒定。

(8) 孔隙比和有效应力之间存在一种与时间无关的唯一关系。

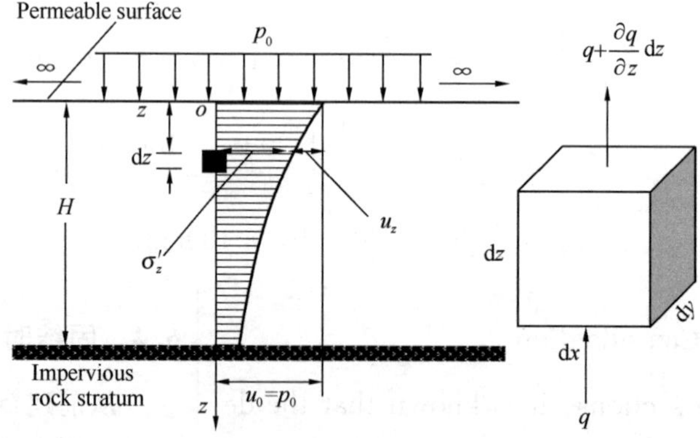

Figure 6.8 Analysis of consolidation process of saturated cohesive soil

As shown in Figure 6.8, the top surface of the saturated soil layer with a thickness of H is permeable, while the bottom surface is impermeable. The consolidation deformation of the soil layer under its own weight has been completed, but the consolidation deformation of the soil layer is only caused by a continuous uniformly distributed load p_0 applied on the permeable surface. The additional stress on the foundation caused by this continuous uniformly distributed load p_0 is uniformly distributed along the depth as $\sigma_z = p_0$, and

如图 6.8 所示，厚度为 H 的饱和土层的顶面透水，而底面不透水。土层在自重作用下的固结变形已经完成，但土层的固结变形仅由施加在透水表面上的连续均匀分布荷载 p_0 引起。由这种连续均匀分布的荷载 p_0 引起的基础上的附加应力沿深度均匀分布，$\sigma_z = p_0$，在 $t=0$ 时完全

it is entirely borne by pore water at time $t=0$. The excess pore water pressure in the soil layer is uniformly distributed along the depth as $u=\sigma_z=p_0$. Due to the impermeability of the lower boundary of the soil layer, pore water flows upward, and the excess pore water pressure at the upper boundary first dissipates completely, while the effective stress begins to increase completely, forming a dissipation curve downwards, namely the growth curve; As time progresses, $t>0$, the excess pore water pressure at a certain point in the layer gradually decreases; And the effective stress gradually increases, while the deformation of the subsoil is growing.

由孔隙水承担。土层中的超孔隙水压力沿深度均匀分布，$u=\sigma_z=p_0$。由于土层下边界的不透水性，孔隙水向上流动，上边界超孔隙水压力首先完全消散，而有效应力开始增大，形成向下的消散曲线，即增长曲线；随着时间的推移，$t>0$，层内某点的超孔隙水压力逐渐减小；随着地基变形的增大，有效应力逐渐增大。

(1) Derivation of one-dimensional consolidation equation

(1) 一维固结方程的推导

Consider an element having dimensions dx, dy and dz within a soil layer of thickness H, as shown in Figure 6.8. An increment of total vertical stress $\Delta\sigma$ is applied to the element. Due to the fact that seepage during consolidation can only occur from bottom to top, the unit seepage volume flowing into and out of the element at a certain time t after the external load is applied once are respectively

厚度为 H 的土层中尺寸为 dx，dy 和 dz 的单元如图 6.8 所示。对单元施加总竖向应力 $\Delta\sigma$ 的增量。由于固结过程中的渗流只能自下而上发生，因此在单个外部荷载后的某个时间 t 流入和流出单元的单位渗流量分别为式 (6.22) 和式 (6.23)。

$$q = v_z A = kiA = k\left(-\frac{\partial h}{\partial z}\right)\mathrm{d}x\mathrm{d}y \tag{6.22}$$

$$q' = q+\frac{\partial q}{\partial z}\mathrm{d}z = v_z A+\frac{\partial vz}{\partial z}A = k\left(-\frac{\partial h}{\partial z}-\frac{\partial^2 h}{\partial z^2}\mathrm{d}z\right)\mathrm{d}x\mathrm{d}y \tag{6.23}$$

where k is permeability coefficient in the z-direction (cm/s); i is hydraulic gradient; h is the excess hydrostatic head at depth z below the permeable surface (cm); A is the cross-sectional area of the element, $A=$ dxdy.

式中，k 是 z 方向上的渗透系数 (cm/s)；i 为水力梯度；h 是渗透表面以下深度 z 处的超静水压头 (cm)；A 是单元的横截面面积，$A=\mathrm{d}x\mathrm{d}y$。

Therefore, the amount of water discharged per unit time of the element is

因此，单元体单位时间的排水量可表示为式 (6.24)。

$$\Delta q = q' - q = -k\frac{\partial^2 h}{\partial z^2}\mathrm{d}x\mathrm{d}y\mathrm{d}z \tag{6.24}$$

The change (decrease) in pore volume within the element is

单元内孔隙体积的变化（减小）为式 (6.25)。

$$\frac{\partial V_w}{\partial t} = -\frac{\partial}{\partial t}\left(\frac{e}{1+e_0}\mathrm{d}x\mathrm{d}y\mathrm{d}z\right) \tag{6.25}$$

where e_0 is natural void ratio of soil; V_w is volume reduction of pore water.

式中，e_0 是土的天然孔隙比；V_w是孔隙水的体积减小量。

According to the continuous condition of consolidation seepage, the change in seepage volume of the element at a certain time t should be equal to the change in pore volume of the element at the same time t. Therefore, it can be derived that

根据固结渗流的连续条件，单元在一定时间 t 的渗流体积变化应等于其在同一时间 t 的孔隙体积变化。因此，可得式(6.26)。

$$\frac{1}{1+e_0}\frac{\partial e}{\partial t}=k\frac{\partial^2 h}{\partial z^2} \tag{6.26}$$

According to the lateral constraint condition of the stress-strain relationship of soil, $de=-adp=-ad\sigma'$, it can be obtained that

根据应力-应变关系的侧限条件，$de=-adp=-ad\sigma'$，可得式（6.27）。

$$\frac{\partial e}{\partial t}=-a\frac{\partial \sigma'}{\partial t} \tag{6.27}$$

where a is compressibility coefficient of soil; σ' is effective stress in soil.

式中，a 是土的压缩系数；σ' 是土中的有效应力。

Substituting Eq. (6.27) into Eq. (6.26) yields

将式(6.27) 代入式(6.26)可得式（6.28）。

$$\frac{k(1+e_0)}{a}\frac{\partial^2 h}{\partial z^2}=-\frac{\partial \sigma'}{\partial t} \tag{6.28}$$

According to the principle of effective stress,

根据有效应力原理，可得式(6.29)。

$$\sigma'=\sigma_z-u \tag{6.29}$$

where σ_z is additional stress in the elemen, $\sigma_z=p$; u is the excess pore water pressure in the element, $u=h\gamma_w$ (where γ_w is the unit weight of water) .

式中，σ_z是单元中的附加应力，$\sigma_z=p$；u 是单元中的超孔隙水压力，$u=h\gamma_w$（其中 γ_w是水的重度）。

Substituting Eq. (6.29) into Eq. (6.28) yields

将式（6.29）代入式(6.28) 得式（6.30）。

$$\frac{\partial u}{\partial t}=\frac{k(1+e_0)}{\gamma_w a}\frac{\partial^2 u}{\partial z^2} \tag{6.30}$$

Let

$$C_v=\frac{k(1+e_0)}{\gamma_w a} \tag{6.31}$$

Eq. (6.30) can be rewritten as follows:

定义固结系数见式（6.31)，则式（6.30）可改写为式（6.32)，

$$\frac{\partial u}{\partial t}=C_v\frac{\partial^2 u}{\partial z^2} \tag{6.32}$$

where C_v is the vertical consolidation coefficient of soil (cm^2/s), it is a function of the permeability coefficient

式中，C_v是土的竖向固结系数(cm^2/s)，它是渗透系数 k、压

k, compression coefficient a, and natural void ratio e_0, and is generally measured directly through consolidation tests. Eq. (6.32) is the famous one-dimensional consolidation equation for saturated soil.

缩系数 a 和天然孔隙比 e_0 的函数，通常通过固结试验直接测量。式（6.32）便是著名的饱和土的一维固结方程。

(2) Analytical solution of one-dimensional consolidation equation

（2）一维固结方程的解析解

When the soil layer is drained on one side, the initial excess pore water pressure is linearly distributed, as shown in Figure 6.9.

当土层一侧排水时，初始超孔隙水压力呈线性分布，如图 6.9 所示。

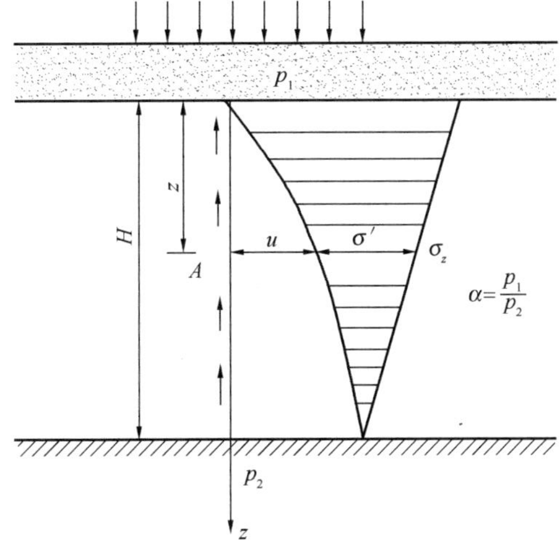

Figure 6.9 Distribution of excess pore water pressure under unidirectional drainage conditions

Assuming the initial excess pore water pressure of the soil drainage surface is p_1 and the initial excess pore water pressure of the impermeable surface is p_2, the ratio of the two is

假设土排水面的初始超孔隙水压力为 p_1，不透水面的初始超孔水压力为 p_2，两者之比为式（6.33）。

$$\alpha = \frac{p_1}{p_2} \tag{6.33}$$

The initial excess pore water pressure u_z at depth z is

深度 z 处的初始超孔隙水压力 u_z 为式（6.34）。

$$u_z = p_2\left[1 + (\alpha - 1)\frac{H - z}{H}\right] \tag{6.34}$$

The initial conditions and boundary conditions for the solution are:

when $t=0$ and $0 \leqslant z \leqslant H$,

$u = p_2\left[1 + (\alpha - 1)\dfrac{H - z}{H}\right]$;

when $0<t<\infty$ and $z=0$, $u=0$;

when $0<t<\infty$ and $z=H$,

初始条件和边界条件为：

当 $t=0$ 且 $0 \leqslant z \leqslant H$ 时，$u = p_2[1 + (\alpha - 1)(H - z)/H]$；

当 $0<t<\infty$ 且 $z=0$ 时，$u=0$；

当 $0<t<\infty$ 且 $z=H$ 时，

$$\frac{\partial u}{\partial z}=0;$$

when $t=\infty$ and $0\leqslant z\leqslant H$, $u=0$.

$\partial u/\partial z=0$；

$t=\infty$ 且 $0\leqslant z\leqslant H$ 时，$u=0$。

According to the above initial conditions and boundary conditions, the special solution of Eq. (6.32) can be obtained by using the separation variable method as follows:

根据以上初始条件和边界条件，采用分离变量法可求得式（6.32）的特解，如式（6.35）所示。

$$u(z,t)=\frac{4p_2}{\pi^2}\sum_{m=1}^{\infty}\frac{1}{m^2}\left[m\pi\alpha+2\,(-1)^{\frac{m-1}{2}}\,(1-\alpha)\right]\exp\left(-\frac{m^2\pi^2}{4}T_v\right)\sin\frac{m\pi z}{2H} \tag{6.35}$$

In practice, the first value is often taken, that is, $m=1$ is obtained

在实用中常取第一项，即取 $m=1$ 得式（6.36）。

$$u=\frac{4p_2}{\pi^2}\left[\alpha(\pi-2)+2\right]\exp\left(-\frac{\pi^2}{4}T_v\right)\sin\frac{\pi z}{2H} \tag{6.36}$$

When the soil layer is drained on two sides, as shown in Figure 6.10.

土层为双面排水时，如图 6.10 所示。

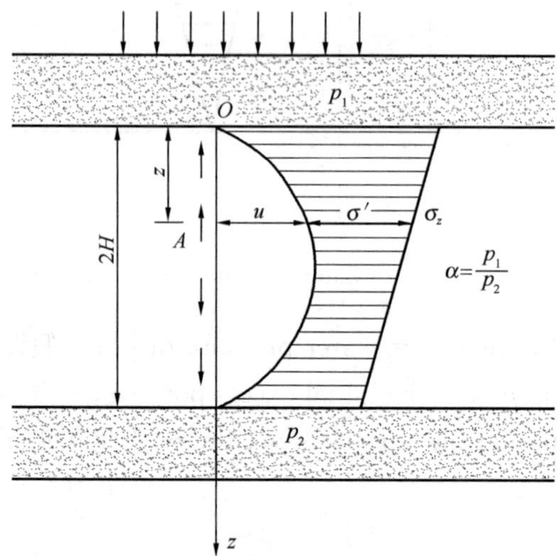

Figure 6.10 Distribution of excess pore water pressure under double-sided drainage conditions

Let $\alpha=p_1/p_2=1$, the thickness of the soil layer is $2H$, and the initial and boundary conditions for solving are:

when $t=0$ and $0\leqslant z\leqslant H$,

$$u=p_2\left[1+(\alpha-1)\frac{H-z}{H}\right];$$

when $0<t<\infty$ and $z=0$, $u=0$;

when $0<t<\infty$ and $z=H$, $u=0$.

令 $\alpha=p_1/p_2=1$，土层厚度为 $2H$，其初始条件和边界条件为：

当 $t=0$ 且 $0\leqslant z\leqslant H$ 时，$u=p_2[1+(\alpha-1)(H-z)/H]$；

当 $0<t<\infty$ 且 $z=0$ 时，$u=0$；

当 $0<t<\infty$ 且 $z=H$ 时，$u=0$。

The special solution of Eq. (6.32) can be obtained by using the separation variable method as follows:

采用分离变量法可求得式(6.32)的特解，如式(6.37)所示。

$$u(z,t)=\frac{p_2}{\pi}\sum_{m=1}^{\infty}\frac{2}{m}[1-(-1)^m\alpha]\exp\left(-\frac{m^2\pi^2}{4}T_v\right)\sin\frac{m\pi(2H-z)}{2H} \tag{6.37}$$

In practice, the first value is often taken, that is, $m=1$ is obtained

在实用中常取第一项，即取 $m=1$ 得式(6.38)。

$$u=\frac{2p_2}{\pi}(1+\alpha)\exp\left(-\frac{\pi^2}{4}T_v\right)\sin\frac{\pi(2H-z)}{2H} \tag{6.38}$$

Where m is positive odd number (1, 3, 5...); exp is exponential functions; H is the farthest drainage distance (cm) of the compressed soil layer, when the soil layer is one-way (above or below) drainage, H takes the thickness of the soil layer, when the double-sided drainage, the water is discharged from the center of the soil layer to the upper and lower directions respectively, and H should take half of the thickness of the soil layer at this time; T_v is the vertical consolidation time factor (dimensionless) and is calculated as follows Eq. (6.39); t is the consolidation duration (s) .

式中，m 为正奇数(1, 3, 5…)；exp 为指数函数；H 为压缩土层最远的排水距离(cm)，当土层为单向(上面或下面)排水时，H 取土层厚度，双面排水时，水由土层中心分别向上下两方向排出，此时 H 应取土层厚度的一半；T_v 为竖向固结时间因数(无量纲)，按式(6.39)计算；t 为固结历时(s)。

$$T_v=C_v t/H^2 \tag{6.39}$$

6.4.2 Degree of Consolidation of Subsoil

(1) The concept of the degree of consolidation

The degree of consolidation refers to the ratio of the consolidation deformation generated by the time t to the final consolidation deformation of the subsoil soil layer under a certain pressure, or the degree of dissipation of the (excess) pore water pressure in the soil layer, also known as the consolidation ratio or consolidation percentage. The definition of foundation consolidation U_z is expressed as follows:

6.4.2 地基固结度

(1) 地基固结度的概念

地基固结度是指地基土层在某一压力作用下，经历时间 t 所产生的固结变形量与最终固结变形量之比值，或土层中超孔隙水压力的消散程度，亦称固结比或固结百分数。地基固结度 U_z 的定义可表达为式(6.40)或式(6.41)。

$$U_z=s_{ct}/s_c \tag{6.40}$$

or

$$U_z=(u_0-u)/u_0 \tag{6.41}$$

where s_{ct} is the amount of foundation consolidation deformation at a certain time t; s_c is the final consolidation deformation of the foundation, and for the normal consolidated soil, the final deformation of the foundation (the final settlement of the foundation) calculated by

式中，s_{ct}是在某一时刻 t 的地基固结变形量；s_c是地基最终固结变形量，对于正常固结土，简化分析取分层总和法单向压缩基本公式计算的地基最终变形量(基

the basic formula of unidirectional compression of the layered sum method is simplified; u_0 is the initial excess pore water pressure (stress) at $t=0$; u is the excess pore water pressure (stress) at time t.

础最终沉降量）；u_0 是 $t=0$ 时的初始超孔隙水压力（应力）；u 是 t 时刻的超孔隙水压力（应力）。

Eq. (6.40) is the strain expression, which is expressed by the ratio of the consolidation deformation at time t to the final consolidation deformation within the depth range of the foundation compression layer at a certain point in the substrate. Eq. (6.41) is a stress expression, which is expressed as the ratio of the effective stress to the total stress at a point in the soil layer. Since the soil is a nonlinear deformation, the results of the two equations are practically not equal.

式（6.40）是应变表达式，即基底某点地基压缩层深度范围内 t 时刻的固结变形量与最终固结变形量之比；而式（6.41）是应力表达式，即土层中某点的有效应力与总应力之比；由于土体为非线性变形体，两式结果实际上是不相等的。

The degree of consolidation at a certain point in the soil layer is not easy to solve the practical engineering problem, so it is necessary to introduce the concept of the average degree of consolidation of a certain soil layer. For vertical drainage, since the consolidation deformation is proportional to the effective stress, the ratio of the effective stress map area to the final effective stress map area at a certain time (Figure 6.8) is called the vertical average consolidation degree $\overline{U}_z$

土层中某点的固结度对于解决实际工程问题并不方便，为此，引入某一土层的平均固结度的概念是必要的。对于竖向排水情况，由于固结变形与有效应力成正比，所以某一时刻有效应力图面积和最终有效应力图面积之比值（图 6.8），称为竖向平均固结度 $\overline{U}_z$，其表达式为式（6.42）。

$$\overline{U}_z = 1-\left(\int_0^H u_{z,t}\mathrm{d}z\right)\Big/\left(\int_0^H \sigma_z \mathrm{d}z\right) \tag{6.42}$$

where $u_{z,t}$ is the excess pore water pressure at a certain time at depth z; σ_z is the vertical additional stress at depth z (generating the initial excess pore water pressure at the time of $t=0$), under the action of continuous uniform load,

式中，$u_{z,t}$ 为深度 z 处某一时刻 t 的超孔隙水压力；σ_z 为深度 z 处的竖向附加应力（产生 $t=0$ 时刻的初始超孔隙水压力），在连续均布荷载 p_0 作用下：

$$\int_0^H \sigma_z \mathrm{d}z = p_0 H \tag{6.43}$$

(2) The average degree of consolidation of the subsoil in the case of an instantaneous application of loads

（2）荷载一次瞬时施加情况的地基平均固结度

When the soil layer is drained from one side, the consolidation degree of the soil layer $\overline{U}_z$ at any time is obtained by substituting Eq. (6.34) and Eq. (6.36) into Eq. (6.43).

土层为单面排水时，将式（6.34）、式（6.36）代入式（6.43）得到任一时刻土层固结度 $\overline{U}_z$，如式（6.44）所示。

$$\overline{U}_z = 1 - \frac{\left(\frac{\pi}{2}\alpha - \alpha + 1\right)}{1+\alpha} \frac{32}{\pi^3} \exp\left(-\frac{\pi^2}{4}T_v\right) \quad (6.44)$$

α takes 1, and the initial excess pore water pressure distribution diagram is rectangular, and its consolidation degree expression is

α 取 1，初始超孔隙水压力分布为矩形，其固结度表达式为式（6.45）。

$$\overline{U}_z = 1 - \frac{8}{\pi^2} \exp\left(-\frac{\pi^2}{4}T_v\right) \quad (6.45)$$

α takes 0, the initial excess pore water pressure distribution diagram is triangular, and its consolidation degree expression is

α 取 0，初始超孔隙水压力分布为三角形，其固结度表达式为式（6.46）。

$$\overline{U}_z = 1 - \frac{32}{\pi^3} \exp\left(-\frac{\pi^2}{4}T_v\right) \quad (6.46)$$

When the soil layer is drained from two sides, the soil consolidation degree $\overline{U}_z$ at any time is obtained by substituting Eq. (6.35) and Eq. (6.39) into Eq. (6.43).

土层为双面排水时，将式（6.35）、式（6.39）代入式（6.43）得任一时刻土层固结度 $\overline{U}_z$，如式（6.47）所示。

$$\overline{U}_z = 1 - \frac{8}{\pi^2} \exp\left(-\frac{\pi^2}{4}T_v\right) \quad (6.47)$$

According to the definition of consolidation degree, the expression for the deformation at any time during the consolidation process can be obtained as

根据固结度的定义，固结过程中任何时刻的变形表达式为式（6.48）。

$$s_{ct} = \overline{U}_z / s_c \quad (6.48)$$

The calculation procedures are as follow:

(1) Calculate the distribution of additional stress along the depth of the soil layer;

(2) Calculate the vertical consolidation deformation of the soil layer;

(3) Calculate the vertical consolidation coefficient and vertical consolidation time factor of the soil layer;

(4) Calculate the (vertical) deformation at a certain moment t during the consolidation process of the soil layer.

其计算步骤如下：

（1）计算附加应力沿土层深度的分布；

（2）计算土层的竖向固结变形；

（3）计算土层的竖向固结系数和垂直固结时间因数；

（4）计算土层固结过程中某一时刻 t 的（竖向）变形。

Example 6.2

Refer to Figure 6.7, Assuming that average consolidation degree of subsoil under foundation $\overline{U}_z = 1 - (8/\pi^2)\exp[-(\pi^2/4)T_v]$. How much is the settlement of the building one year after its completion?

例题 6.2

参考图 6.7，假设地基下地基的平均固结度 $\overline{U}_z = 1 - (8/\pi^2)\exp[-(\pi^2/4)T_v]$。则建筑物竣工一年后的沉降是多少？

Solution

Consolidation coefficient of the subsoil is

解答

地基土的固结系数为

$$C_v = k(1+e_0)/(\gamma_w a) = 5\times10^{-8}\times(1+0.5)/(10^{-5}\times5) = 0.0015\ \text{cm}^2/\text{s}$$

Vertical consolidation time factor is

竖向固结时间因数为

$$T_v = C_v t/H^2 = \frac{0.0015\times1\times365\times24\times3600}{4\times10^2\times4\times10^2} \approx 0.296$$

One year after the completion of the building, the average consolidation degree of the subsoil is

建筑物竣工一年后，地基土的平均固结度为

$$U_z = 1-\frac{8}{\pi^2}\exp(-\pi^2/4T_v) \approx 0.61$$

Therefore, the settlement of the building one year after its completion is

因此，该建筑竣工一年后的沉降量为

$$s_c = 100\times0.61 = 61\text{mm}$$

6.5 Secondary Consolidation Settlement

Secondary consolidation settlement is believed to be related to the skeleton creep of soil. It is a compression that slowly increases over time after the excess pore water pressure has dissipated and the effective stress growth remains basically unchanged. In the process of secondary consolidation settlement, the rate of volume change of soil is independent of the outflow rate of pore water from the soil, that is, the time of secondary consolidation settlement is independent of the thickness of the soil layer. Secondary consolidation settlement is not important compared to consolidation settlement, but for soft clay, especially when the soil contains some organic matter (such as colloidal humus), or when the pressure increment ratio (referring to the ratio of additional stress to self-weight stress in the soil) is small in deep compressible soil layers, secondary consolidation settlement must be taken it into account.

The secondary consolidation settlement of unidirectional compression of the subsoil can be calculated according to the following equation:

6.5 次固结沉降

次固结沉降被认为与土的骨架蠕变有关。这是一种在超孔隙水压力消散后有效应力增长基本保持不变的压缩，随时间缓慢增长。在次固结沉降过程中，土体的体积变化率与孔隙水从土体中流出的速率无关，也就是说，次固结沉降的时间与土层厚度无关。与固结沉降相比，次固结沉降不太重要，但对于软黏土，特别是当土体含有一些有机物（如胶体腐殖质）时，或者当深层可压缩土层中的压力增量比（指土中附加应力与自重应力的比值）较小时，必须考虑次固结沉降。

地基土单向压缩的次固结沉降可根据式（6.49）计算，

$$s_s = \sum_{i=1}^{n}\frac{H_i}{1+e_{0i}}C_{\alpha i}\lg\frac{t}{t_1} \tag{6.49}$$

where s_s is secondary consolidation settlement of the subsoil; H_i is thickness of the i-th soil layer; e_{0i} is void

式中，s_s是地基土的次固结沉降；H_i是第 i 层土的厚度；e_{0i}为

ratio of the i-th soil layer; $C_{\alpha i}$ is secondary consolidation coefficient of the i-th soil layer; t is the time required for the secondary consolidation settlement, $t > t_1$; t_1 is equivalent to the time when the primary consolidation degree is 100%, extrapolated from the e-lg$t$ curve.

第 i 层土的孔隙比；$C_{\alpha i}$ 为第 i 层土的次固结系数；t 为次固结沉降的时间，$t > t_1$；t_1 为主固结度为 100% 的时刻，可从 e-lg$t$ 曲线外推得出。

Problems

6.1 A saturated clay layer with a thickness of 10.0m, consisting of sand layers above and below, as shown in the figure. The additional stress distribution caused by the base pressure in the clay layer is also shown in the Figure 6.11.

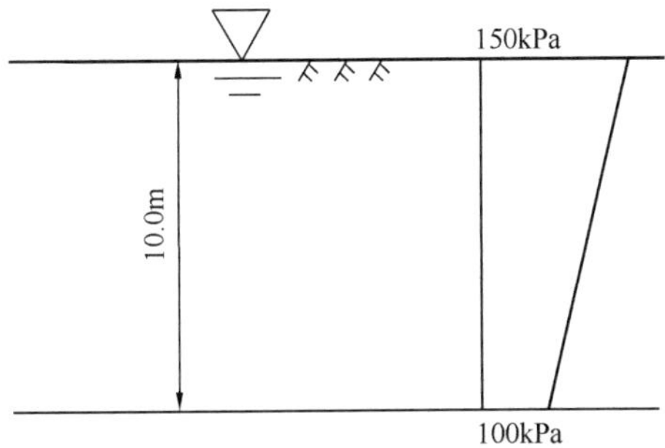

Figure 6.11 Figure of problem 6.1

The physical and mechanical properties of the clay layer have been measured, including initial porosity $e_0 = 0.8$, compression coefficient $a = 0.25\text{MPa}^{-1}$, and permeability coefficient $k = 2.0\text{cm/a}$. Calculate the settlement amount for one year of loading and the time required for the settlement to reach half of the final settlement amount.

6.2 The thickness of a compressed soil layer is 4.0m, and the distribution of additional stress is a function of depth z: $f(z) = 100 - 10z$. The compressibility coefficient of the soil layer was measured to be $a = 0.9\text{MPa}^{-1}$, and the natural porosity ratio was $e_0 = 0.8$. Attempt to determine the final settlement of the soil layer.

Chapter 7 Shear Strength of Soil

第 7 章　土的抗剪强度

7.1 Introduction

When the soil mass is subjected to external forces, it will generate shear stress and shear deformation. The soil mass has the potential ability to resist shear stress—shear resistance, which gradually comes into play with the increase of shear stress. When the shear resistance is fully exerted, the soil is in the limit state of shear failure, and the shear stress reaches its limit, which is the shear strength of the soil. The shear strength of soil can be defined as the ultimate value at which the soil mass resists shear stress, or the ability of the soil mass to resist shear failure. One must understand the nature of shearing resistance in order to analyze soil stability problems, such as bearing capacity, slope stability, and lateral pressure on earth-retaining structures. The following will be introduced in this chapter:

(1) Theory of shear strength of soil;

(2) Laboratory testing of the shear strength of soil.

7.1 概述

当土体受到外力时，会产生剪应力和剪切变形。土体具有抵抗剪应力的潜在能力——剪阻力，随着剪切应力的增大逐渐发挥作用。当剪阻力完全发挥时，土体处于剪切破坏的极限状态，剪应力达到其极限，即土的抗剪强度。土的抗剪强度可以定义为土体抵抗剪应力的极限值，或土体抵抗剪切破坏的能力。为了分析土稳定性，如承载力、边坡稳定性和挡土结构上的侧向压力，必须理解剪阻力的本质。本章将介绍以下内容：

(1) 土的抗剪强度理论；

(2) 土的抗剪强度的实验室测试。

7.2 Theory of Shear Strength of Soil

In 1773, Coulomb expressed the shear strength τ_f of soil as a function of the total normal stress σ on the shear failure surface based on direct shear tests of sandy soil, that is

7.2 土体抗剪强度理论

1773 年，库仑基于砂土的直接剪切试验，将土的抗剪强度 τ_f 表达为剪切破坏面上法向总应力 σ 的函数，即式 (7.1)。

$$\tau_f = \sigma \tan \varphi \tag{7.1}$$

Subsequently, a more general expression suitable for cohesive soil was proposed:

后来又提出了适合黏性土的更通用的表达式 (7.2)。

$$\tau_f = c + \sigma \tan\varphi \tag{7.2}$$

where τ_f is the shear strength (kPa); σ is normal stress on the failure plane; c is cohesion (kPa); φ is angle of internal friction (°).

式中，τ_f 为抗剪强度 (kPa)；σ 为破坏面上的总应力；c 为黏聚力 (kPa)；φ 为内摩擦角 (°)。

Eqs. (7.1) and (7.2) are referred to as Coulomb's

式 (7.1) 和式 (7.2) 称为

formula or Coulomb's law, where c and φ are referred to as shear strength parameters. The Coulomb formula in the σ-τ_f coordinate system is represented by two straight lines, as shown in Figure 7.1. It can be called Coulomb strength line. According to Coulomb's formula, the shear strength of cohesionless soil is directly proportional to the normal stress on the shear plane and is also related to the internal friction angle. The shear strength of cohesionless soil mainly depends on factors such as the roughness of the soil particle surface, the compactness of the soil, and the particle size distribution. The shear strength of cohesive soil and loess consists of two parts, one is frictional resistance (proportional to normal stress); The other part is the cohesive force that is independent of normal stress and resists the sliding between soil particles, mainly caused by factors such as bonding between clay particles and electrostatic attraction.

库仑公式或库仑定律，其中 c 和 φ 称为抗剪强度指标。σ-τ_f 坐标系中库仑公式由两条直线表示，如图 7.1 所示。可称之为库仑强度线。由库仑公式可知，无黏性土的抗剪强度与剪切面上的法向应力成正比，还与内摩擦角有关，无黏性土的抗剪强度主要取决于土粒表面的粗糙度、土体的密实度以及颗粒级配等因素。黏性土和粉土的抗剪强度由两部分组成，一部分是摩阻力（与法向应力成正比）；另一部分是与法向应力无关的、抵抗土体颗粒间相互滑动的黏聚力，主要由黏土颗粒之间的胶结作用和静电引力效应等因素引起。

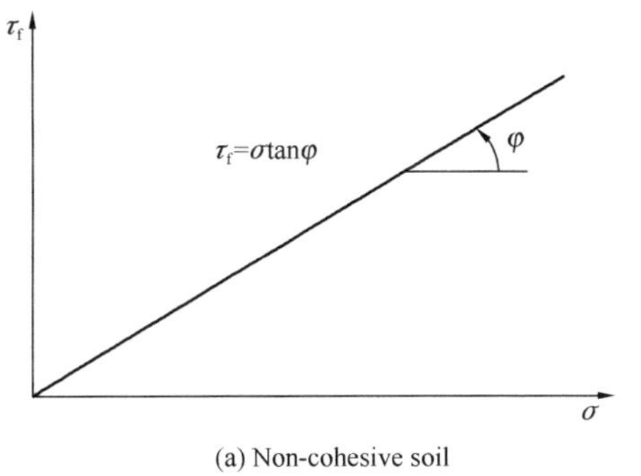

(a) Non-cohesive soil

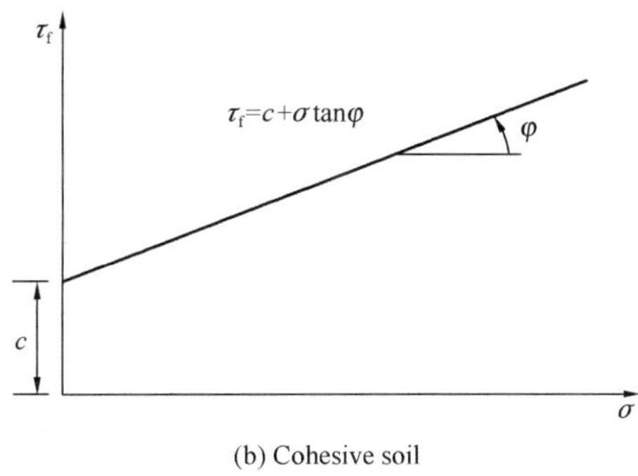

(b) Cohesive soil

Figure 7.1 Coulomb strength line

Coulomb's law can also be expressed as effective stress:

库仑定律也可以以有效应力表示为式（7.3）。

$$\tau_f = c' + \sigma' \tan\varphi' \tag{7.3}$$

where σ' is effective normal stress on the failure plane; c' is effective cohesion (kPa); φ' is effective angle of internal friction (°).

式中，σ'为剪切破坏面上的有效法向应力；c' 为有效黏聚力（kPa）；φ'为有效内摩擦角（°）。

Mohr (1900) presented a theory for rupture in materials that contended that a material fails because of a critical combination of normal stress and shearing stress

莫尔于 1900 年提出了一种材料破坏理论，该理论认为材料破坏是正应力与剪应力的临界组合所致，而不是仅是最大正应力

and not from either maximum normal or shear stress alone. And express the functional relationship between the normal stress and shear stress on the failure surface as:

或剪应力的单独作用。并将破坏面上的正应力和剪应力之间的函数关系表示为式（7.4）。

$$\tau_f = f(\sigma) \tag{7.4}$$

The failure envelope defined by Eq. (7.4) is a curved line. Therefore, Coulomb′s law is a special case of Eq. (7.4). In fact, for most soil mechanics problems, it is sufficient to approximate the shear stress on the failure plane as a linear function of the normal stress (Coulomb′s law).

由式（7.4）定义的破坏包络线是一条曲线。也就是说，库仑定律为式（7.4）的一个特例。事实上，对于大多数土力学问题，将破坏面上的剪应力近似为正应力的线性函数（库仑定律）就足够了。

7.2.1 The Stress State of a Point in the Soil

7.2.1 土中某点的应力状态

As stated by the Mohr-Coulomb failure criterion, failure from shear will occur when the shear stress on a plane reaches a value given by Eq. (7.2). Taking the plane strain problem as an example.

根据莫尔-库仑破坏准则，当平面上的剪应力达到式（7.2）所得值时，将会发生剪切破坏。

Take a micro element in the soil (Figure 7.2a), and assume that the two principal stresses acting on the element are σ_1 and σ_3 ($\sigma_1 > \sigma_3$). Within the element, there are normal stresses σ and shear stresses τ on the *mn* plane at any angle α to the plane of action of the major principal stress σ_1. In order to establish the relationship between σ, τ, and σ_1, σ_3, the micro prism is taken as the isolator (Figure 7.2b), and each force is projected in the horizontal and vertical directions respectively. Based on the static equilibrium conditions, yields

以平面应变问题为例，在土中取一微单元体（图 7.2a），设作用在该单元体上的两个主应力为 σ_1 和 σ_3（$\sigma_1 > \sigma_3$）。在单元体内与大主应力 σ_1 作用平面成任意角 α 的 *mn* 平面上有正应力 σ 和剪应力 τ。为建立 σ、τ 与 σ_1、σ_3 之间的关系，取微棱柱体为隔离体（图 7.2b），将各力分别在水平和垂直方向投影，根据静力平衡条件得式（7.5）与式（7.6）。

$$\sigma_3 \mathrm{d}s\sin\alpha - \sigma \mathrm{d}s\sin\alpha + \tau \mathrm{d}s\cos\alpha = 0 \tag{7.5}$$

$$\sigma_1 \mathrm{d}s\cos\alpha - \sigma \mathrm{d}s\cos\alpha - \tau \mathrm{d}s\sin\alpha = 0 \tag{7.6}$$

From Eqs. (7.5) and (7.6), the normal stress and shear stress on the *mn* plane can be derived as

根据式（7.5）和式（7.6），*mn* 平面上的正应力和剪应力可以推导式（7.7）和式（7.8）。

$$\sigma = \frac{1}{2}(\sigma_1 + \sigma_3) + \frac{1}{2}(\sigma_1 - \sigma_3)\cos 2\alpha \tag{7.7}$$

$$\tau = \frac{1}{2}(\sigma_1 - \sigma_3)\sin 2\alpha \tag{7.8}$$

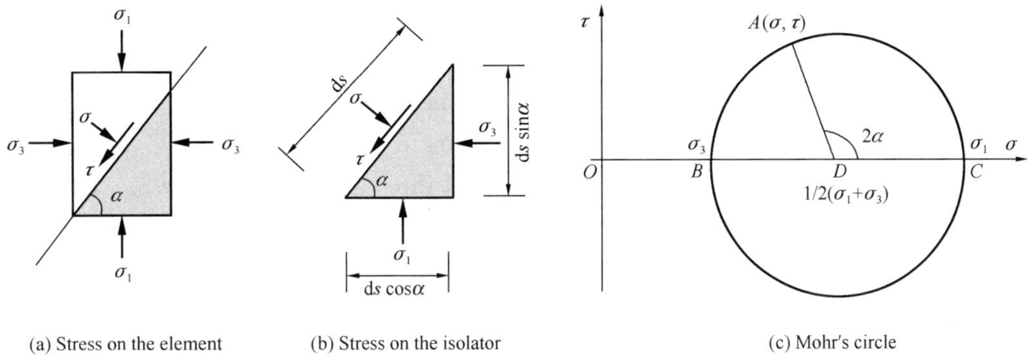

(a) Stress on the element (b) Stress on the isolator (c) Mohr's circle

Figure 7.2 Stress at any point in the soil

From Eqs. (7.7) and (7.8), it can be obtained that

由式（7.7）和式（7.8），可得式（7.9）。

$$\left[\sigma - \frac{1}{2}(\sigma_1 + \sigma_3)\right]^2 + \tau^2 = \left[\frac{1}{2}(\sigma_1 - \sigma_3)\right]^2 \tag{7.9}$$

Therefore, Eq. (7.9) represents a circle in the σ-τ Cartesian coordinate system with [$(\sigma_1+\sigma_3)/2$, 0] as the center and $(\sigma_1-\sigma_3)/2$ as the radius. This circle is called the Mohr's circle (Figure 7.2c) . Starting from *DC*, rotate counter clockwise by an angle of 2α so that the *DA* line intersects with the circumference at point *A*. It can be proven that the horizontal axis of point *A* is the normal stress σ on the inclined surface *mn*, and the vertical axis is the shear stress τ. In this way, the Mohr's circle can represent the stress state of a point in the soil, and the coordinates of each point on the circumference of the Mohr's circle represent the normal stress and shear stress of that point on the corresponding plane.

因此，式（7.9）表示 σ-τ 直角坐标系中的一个圆，圆心坐标为 [$(\sigma_1+\sigma_3)/2$, 0]，半径为 $(\sigma_1-\sigma_3)/2$。这个圆被称为莫尔圆（图 7.2c）。从 *DC* 开始逆时针旋转 2α 角，使 *DA* 线与圆周交于 *A* 点。可以证明，*A* 点的横坐标即为斜面 *mn* 上的正应力 σ，纵坐标即为剪应力 τ。这样，莫尔圆就可以表示土体中一点的应力状态，莫尔圆圆周上各点的坐标就表示该点在相应平面上的正应力和剪应力。

If the shear strength parameters c and φ of the soil and the stress state at a certain point in the soil are given, the shear strength envelope can be plotted on the same coordinate graph as the Mohr's circle (Figure 7.3). There are three possible relationships between them: ① The entire Mohr's circle (circle Ⅰ) is located below the shear strength envelope, indicating that the shear stress at this point in any plane is less than the shear strength that the soil can exert ($\tau < \tau_f$), and

如果给定了土体的抗剪强度参数 c、φ 以及土中某点的应力状态，则可将抗剪强度包线与莫尔圆绘制在同一张坐标图上，如图 7.3 所示。它们之间的关系有以下三种情况：①整个莫尔圆（圆Ⅰ）位于抗剪强度包线下方，说明该点在任何平面上的剪应力都小于土体所能发挥的抗剪强度

therefore no shear failure will occur; ② The Mohr's circle (circle Ⅱ) is tangent to the shear strength envelope at point A, indicating that on the plane represented by point A, the shear stress is exactly equal to the shear strength ($\tau=\tau_f$), and this point is in a state of ultimate equilibrium. This Mohr's circle (circle Ⅱ) is called the limiting stress circle; ③ The shear strength envelope is a tangent line of the Mohr's circle (circle Ⅲ, represented by a dashed line). In fact, this situation cannot exist because the shear stress in any direction at this point cannot exceed the shear strength of the soil, that is, there is no situation where $\tau > \tau_f$.

（$\tau<\tau_f$），因此不会发生剪切破坏；②莫尔圆（圆Ⅱ）与抗剪强度包线相切，切点为A，表明在点A所代表的平面上，剪应力恰好等于抗剪强度（$\tau=\tau_f$），该点处于极限平衡状态，此莫尔圆称为极限应力圆；③抗剪强度包线是莫尔圆（圆Ⅲ，以虚线表示）的一条割线，实际上这种情况不可能存在，因为该点任何方向上的剪应力都不可能超过土体的抗剪强度，即不存在$\tau>\tau_f$的情况。

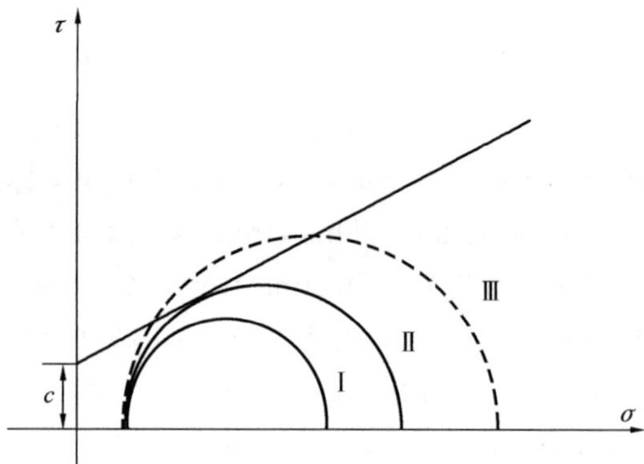

Figure 7.3 The relationship between Mohr's circle and strength line

7.2.2 Limit Equilibrium Condition

Based on the geometric relationship between the limit Mohr's circle and the Coulomb strength line, the limit equilibrium condition can be derived.

Take a micro unit in the soil, as shown in Figure 7.4(a), where mn is the fracture surface, which forms a fracture angle α_f with the surface of the major principal stress. The Mohr circle at this point in the limit equilibrium state is shown in Figure 7.4 (b). Extend the shear strength envelope and intersect the σ axis at point R. From the triangle ARD, it can be obtained that $\overline{AD}=\overline{RD}\sin\varphi$.

$\overline{AD}=1/2(\sigma_1-\sigma_3)$, $\overline{RD}=c\cot\varphi+1/2(\sigma_1+\sigma_3)$, therefore, $\sin\varphi=(\sigma_1-\sigma_3)/(\sigma_1+\sigma_3+2c\cot\varphi)$.

7.2.2 极限平衡条件

根据极限莫尔应力圆和库仑强度线之间的几何关系，可推导出极限平衡条件。

在土体中取一微单元体，如图7.4（a）所示，其中mn为破裂面，与大主应力的作用面呈破裂角α_f。该点处于极限平衡状态时的莫尔圆如图7.4（b）所示。将抗剪强度包线延长与σ轴相交于R点，由三角形ARD可知：$\overline{AD}=\overline{RD}\sin\varphi$。

因$\overline{AD}=1/2(\sigma_1-\sigma_3)$，$\overline{RD}=c\cot\varphi+1/2(\sigma_1+\sigma_3)$，可得$\sin\varphi=(\sigma_1-\sigma_3)/(\sigma_1+\sigma_3+2c\cot\varphi)$。

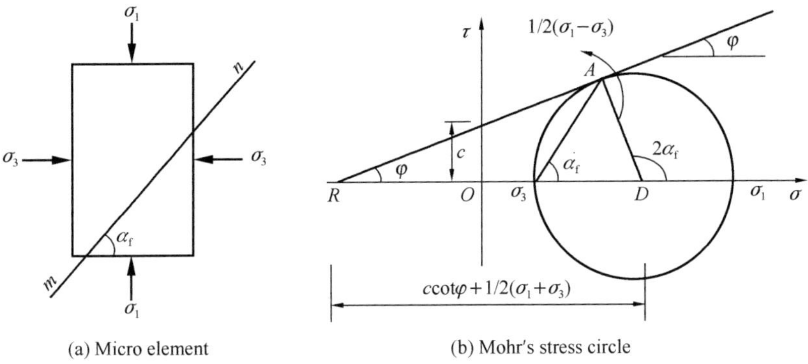

(a) Micro element (b) Mohr's stress circle

Figure 7.4 Mohr's stress circle under extreme state

Therefore,

化简后可得式（7.10）或式（7.11）。

$$\sigma_1 = \sigma_3 \frac{1+\sin\varphi}{1-\sin\varphi} + 2c\sqrt{\frac{1+\sin\varphi}{1-\sin\varphi}} \tag{7.10}$$

or

$$\sigma_3 = \sigma_1 \frac{1-\sin\varphi}{1+\sin\varphi} - 2c\sqrt{\frac{1-\sin\varphi}{1+\sin\varphi}} \tag{7.11}$$

It can be proven by trigonometric functions

由三角函数可证明式（7.12）和式（7.13）成立。

$$\frac{1+\sin\varphi}{1-\sin\varphi} = \tan^2(45° + \frac{\varphi}{2}) \tag{7.12}$$

or

$$\frac{1-\sin\varphi}{1+\sin\varphi} = \tan^2(45° - \frac{\varphi}{2}) \tag{7.13}$$

Substituting Eqs. (7.12) and (7.13) into Eqs. (7.10) and (7.11), respectively, yields

将式（7.12）和式（7.13）分别代入式（7.10）和式（7.11），得到式（7.14）和式（7.15）

$$\sigma_1 = \sigma_3 \tan^2(45° + \frac{\varphi}{2}) + 2c\tan(45° + \frac{\varphi}{2}) \tag{7.14}$$

or

$$\sigma_3 = \sigma_1 \tan^2(45° - \frac{\varphi}{2}) - 2c\tan(45° - \frac{\varphi}{2}) \tag{7.15}$$

For non-cohesive soil, $c=0$, therefore

对于无黏性土，$c=0$，可得式（7.16）和式（7.17）。

$$\sigma_1 = \sigma_3 \tan^2(45° + \frac{\varphi}{2}) \tag{7.16}$$

or

$$\sigma_3 = \sigma_1 \tan^2(45° - \frac{\varphi}{2}) \tag{7.17}$$

In the triangle *ARD* of Figure 7.4 (b), the rupture angle can be obtained from the relationship between the outer and inner angles that $\alpha_f = 45° + \varphi/2$. The angle between the failure surface and the surface of the major principal stress σ_1 is $(45° + \varphi/2)$, or the angle between the failure surface and the surface of the small principal stress σ_3 is $(45° - \varphi/2)$.

在图 7.4（b）的三角形 *ARD* 中，破裂角可由外角和内角之间的关系 $\alpha_f = 45° + \varphi/2$ 所得。破坏面与大主应力 σ_1 作用面之间的夹角为 $(45° + \varphi/2)$，或破坏面与小主应力 σ_3 作用面之间的夹角为 $(45° - \varphi/2)$。

7.3 Laboratory Test for Determination of Shear Strength Parameters

There are several laboratory methods available to determine the shear strength parameters (i.e., c, φ, c', φ') of various soil samples in the laboratory. They are as follows:

(1) Direct shear test;

(2) Triaxial test;

(3) Unconfined compressive strength test.

The direct shear test and the triaxial test are the two commonly used techniques for determining the shear strength parameters. These two tests will be described in detail in the sections that follow.

7.3 测定抗剪强度参数的室内试验

有几种试验方法可用于确定各种土试样的抗剪强度参数（即 c，φ，c'，φ'），分别是：

（1）直接剪切试验；

（2）三轴试验；

（3）无侧限抗压强度试验。

室内的直接剪切试验和三轴试验是确定抗剪强度参数的两种常用方法。下面将详细介绍这两种试验。

7.3.1 Direct Shear Test

Test procedures are detailed in "Standard for Geotechnical Testing Method" GB/T 50123—2019. The sample, of cross-sectional area A, is confined in a metal box (known as the shear box or direct shear apparatus) of square or circular cross-section split horizontally at mid-height, a small clearance being maintained between the two halves of the box. At failure within an element of soil under principal stresses σ_1 and σ_3, a slip plane will form within the element at an angle α as shown in Figure 7.2. The shear box is designed to represent the stress conditions along this slip plane. Porous plates are placed below and on top of the specimen if it is fully or partially saturated to allow free drainage: if the sample is dry, solid metal plates may be used. The essential features of the apparatus are

7.3.1 直接剪切试验

试验步骤详见《土工试验方法标准》GB/T 50123—2019。横截面积为 A 的试样被放置在一个方形或圆形横截面的金属箱（称为剪切箱或直剪仪）中，在该金属箱中间高度水平分割，箱的上下两半之间有一个小间隙。在主应力 σ_1 和 σ_3 作用下，土单元体内发生破坏时，单元体内将形成一个角度为 α 的滑移面，如图 7.2 所示。剪切箱旨在表示沿该滑动面的应力条件。如果试样完全或部分饱和，则将多孔板放置在试样下方和上方，以允许自由排水：如果试样干燥，则可

shown in Figure 7.5. In Figure 7.5, a vertical force (N) is applied to the sample through a loading plate, under which the sample is allowed to consolidate. Shear displacement is then gradually applied on a horizontal plane by causing the two halves of the box to move relative to each other, the shear force required (T) being measured together with the corresponding shear displacement (Δl). The induced shear stress within the sample on the slip plane is equal to that required to shear the two halves of the box. Normally, the change in thickness (Δh) of the sample is also measured. If the initial thickness of the sample is h_0, then the shear strain (γ) can be approximated by $\Delta l/h_0$ and the volumetric strain (ε_v) by $\Delta h/h_0$.

以使用固体金属板。该装置的基本特征如图7.5所示。在图7.5中，通过加载板对试样施加垂直力N，使试样在加载板下固结。然后，通过使剪切盒的上下两半相对移动，在水平面上逐渐施加剪切位移，所需的剪切力T与相应的剪切位移Δl一起测量。滑动面上试样内的剪应力等于剪切盒两半所需的剪应力。通常，还测量试样厚度的变化Δh。如果试样的初始厚度为h_0，则剪切应变γ可以用$\Delta l/h_0$近似，体积应变ε_v可以用$\Delta h/h_0$近似。

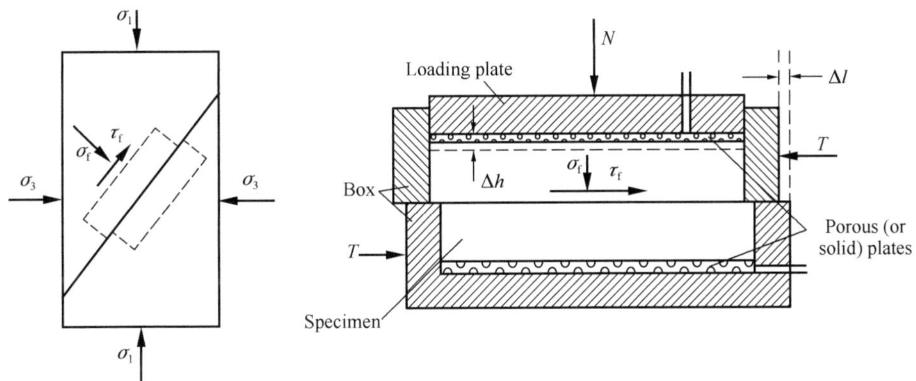

Figure 7.5 Schematic direct shear apparatus

A number of sample of the soil are tested, each under a different vertical force, and the value of shear stress at failure ($\tau_f = T/A$) is plotted against the normal effective stress ($\sigma_f = N/A$) for each test. The Mohr-Coulomb shear strength parameters c and φ are then obtained from the straight line which best fits the plotted points (Figure 7.6). The shear stress throughout the test may also be plotted against the shear strain; the gradient of the initial part of the curve before failure (peak shear stress) gives a crude approximation of the shear modulus G.

对多个土试样进行测试，每个试样受到不同的竖向力，并绘制每次试验破坏时的剪应力值（$\tau_f = T/A$）与法向有效应力（$\sigma_f = N/A$）的关系图。然后，从各点连成的最接近的直线上获得莫尔-库仑抗剪强度参数c和φ（图7.6）。也可以绘制试验过程中剪应力和剪应变的关系曲线，破坏前曲线初始部分的梯度（峰值剪切应力）可给出剪切模量G的近似值。

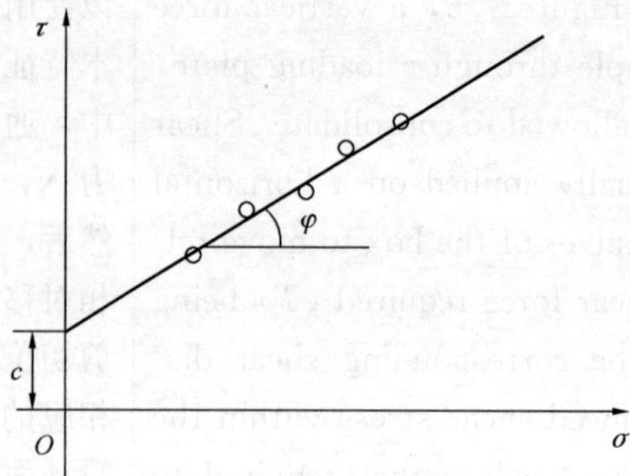

Figure 7.6 Determination of the values of c and φ from direct shear test

7.3.2 Triaxial Test

The triaxial apparatus is the most widely used laboratory device for measuring soil behavior in shear, and is suitable for all types of soil. The test has the advantages that drainage conditions can be controlled, and pore water pressure measurements can be made. A cylindrical sample, generally having a length/diameter ratio of 2, is used in the test; this sits within a chamber of pressurized water. The sample is stressed axially by a loading ram and radially by the confining fluid pressure under conditions of axial symmetry in the manner shown in Figure 7.7. The most common test, triaxial compression, involves applying shear to the soil by holding the confining pressure constant and applying compressive axial load through the loading ram.

7.3.2 三轴试验

三轴仪是测量土剪切行为最常用的试验设备，适用于所有类型的土。该试验的优点是可以控制排水条件，并可以进行孔隙水压力测量。试验中使用圆柱形试样，其长度/直径通常为 2；它位于一个加压的水腔内。在轴对称条件下，试样受到加载板的轴向应力和围压流体的径向应力，如图 7.7 所示。最常见的三轴压缩试验是在围压恒定的情况下通过加载板施加压缩轴向荷载来对土施加剪切力。

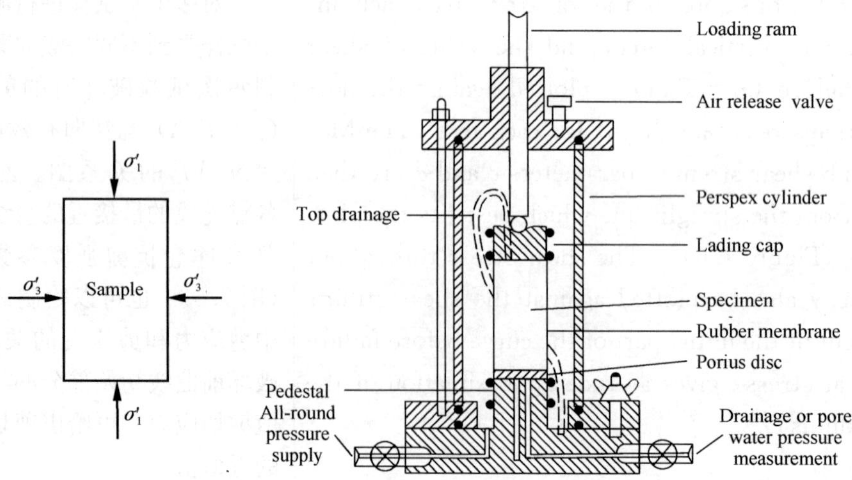

Figure 7.7 Schematic triaxial apparatus

The main features of the apparatus are also shown in Figure 7.7. The circular base has a central pedestal on which the sample is placed, there being access through the pedestal for drainage and for the measurement of pore water pressure. The sample is placed on either a porous or a solid disc on the pedestal of the apparatus and sealed in a cylindrical organic glass cylinder. Typical sample diameters are 38mm and 100mm. A small negative pressure is applied to the pore water to maintain the stability of the sample while the negative pressure is removed prior to the application of the all-round confining pressure. The sample is subjected to an all-round fluid pressure σ_3 in the cell.

At this point, the three principal stresses in all directions within the sample are equal, so no shear stress is generated (Figure 7.8a). Then, vertical pressure is applied to the sample through a force transfer rod, so that the vertical principal stress is greater than the horizontal principal stress. When the horizontal principal stress remains constant and the vertical principal stress gradually increases, the sample is ultimately sheared and destroyed (Figure 7.8b). Assuming that the increment of vertical compressive stress applied to the sample by the force transfer rod during shear failure is $\Delta\sigma_1$, the major principal stress on the sample is $\sigma_1=\sigma_3+\Delta\sigma_1$, and the small principal stress is σ_3. A limit stress circle can be drawn with ($\sigma_1-\sigma_3$) as the diameter, as shown in circle A in Figure 7.8 (c). Several samples (3 or more) of the same soil sample were tested using the above method, and different confining pressures (σ_3) were applied to each sample to obtain the maximum principal stress (σ_1) at shear failure. These results were plotted as a set of ultimate stress circles, as shown in circles A, B, and C in Figure 7.8 (c). Due to the fact that these samples were all sheared to failure, according to the Mohr-Coulomb theory, a set of common tangents to the ultimate stress circle are drawn as the shear strength envelope of the soil, which is usually approximated as a straight line. The angle

该装置的主要特征如图 7.7 所示。圆形基座中央有一个底座，试样放置在基座上，可以通过基座排水和测量孔隙水压力。将试样放置在设备底座上的多孔或实心圆盘上，并密封在有机玻璃圆柱体中。典型的试样直径为 38mm 和 100mm。对孔隙水施加一个小的负压，以保持试样的稳定性，同时在施加全方位围压之前去除负压。试样在围压室中承受各向的流体压力 σ_3。

此时，试样内各方向上的三个主应力相等，因此不会产生剪切应力（图 7.8a)。然后，通过传力杆对试样施加竖向压力，使竖向主应力大于水平主应力。当水平主应力保持恒定，竖向主应力逐渐增大时，试样最终被剪切破坏（图 7.8b)。假设在剪切破坏过程中，传力杆施加在试样上的竖向压应力增量为 $\Delta\sigma_1$，则试样上的大主应力为 $\sigma_1=\sigma_3+\Delta\sigma_1$；小主应力则为 σ_3。以（$\sigma_1-\sigma_3$）为直径可绘出一个极限应力圆，如图 7.8（c）所示圆 A。使用上述方法对同一土样的若干试样（3 个及 3 个以上）进行测试，并对每个试样施加不同的围压（σ_3)，可以得出剪切破坏时的大主应力（σ_1)。这些结果可绘制为一组极限应力圆，即图 7.8（c）中的圆 A、B、C。由于这些试样都剪切至破坏，根据莫尔-库仑理论，绘制一组极限应力圆的公共切线作为土的抗剪强度包线，通常近似为一条直线。该直线与横坐标之间的夹角为土的内摩擦角 φ，直线与纵坐标的截距

between this line and the horizontal axis is the internal friction angle φ of the soil, and the intercept between the line and the vertical axis is the soil's cohesive force, c.

为土的黏聚力 c。

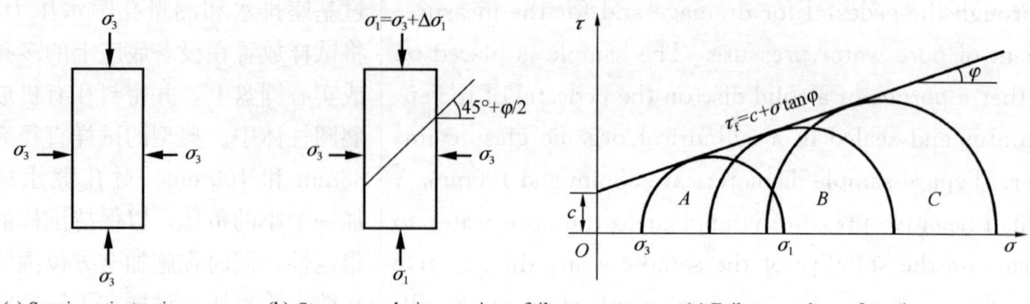

Figure 7.8 Principle of triaxial compression test

If measuring the pore water pressure during the testing process, the pore water pressure valve can be opened. After applying pressure on the sample, the increase in pore water pressure in the soil forces the mercury level of the zero-position indicator to drop. To measure the pore water pressure, the mercury level of the zero-position indicator can be adjusted using a pressure regulating cylinder to maintain its original position. In this way, the reading on the pore water pressure gauge is the pore water pressure value. To measure the displacement during the test, the drainage valve can be opened to allow the water in the sample to be discharged into the measuring pipe. The displacement during the test can be calculated based on the change in water level in the measuring pipe.

如果在测试过程中测量孔隙水压力，可以打开孔隙水压力阀。在试样上施加压力后，土中孔隙水压力的增大迫使零位指示器的水银面下降。为测量孔隙水压力，可用压力调节筒调整零位指示器的水银面保持其原始位置。这样，孔隙水压力表上的读数就是孔隙水压力值。如需测量试验过程中排水量，可以打开排水阀，将试样中的水排入测量管。试验过程中的排水量可以根据测量管中水位的变化来计算。

There are three methods for triaxial compression testing based on the consolidation state of the surrounding pressure σ_3 before shear and the drainage conditions during shear:

根据剪切前周围压力 σ_3 的固结状态和剪切过程中的排水条件，三轴压缩试验有三种方法：

(1) Unconsolidated-Undrained (UU). The specimen is subjected to a specified confining pressure and then the principal stress difference is applied immediately, with no drainage/consolidation being permitted at any stage of the test.

（1）不固结不排水三轴试验（UU）。试样承受恒定的围压，然后立即施加主应力差，在试验的任何阶段都不允许排水/固结。

(2) Consolidated-Undrained (CU). The sample is subjected to confining pressure σ_3 and the drainage

（2）固结不排水三轴试验（CU）。试样承受围压 σ_3，打开

valve is opened to allow drainage consolidation. After consolidation is stable, the drainage valve is closed and vertical pressure is applied to shear failure of the sample under undrained conditions.

(3) Consolidated-Drained (CD) . Drainage of the sample is permitted under a specified confining pressure until consolidation is complete; with drainage still being permitted, the principal stress difference is then applied at a rate slow enough to ensure that the excess pore water pressure is maintained at zero.

The test procedure of the above three principal types of test is standardized in " Standard for Geotechnical Testing Methods" GB/T 50123—2019.

排水阀进行排水固结。固结稳定后，关闭排水阀，在不排水条件下对试样施加竖向压力直至剪切破坏。

(3) 固结排水三轴试验(CD)。允许试样在规定的围压下排水，直至固结完成；在允许排水的情况下，以足够慢的速度施加主应力差，以确保多余的孔隙水压力为零。

上述三种基本试验的试验方法可参考《土工试验方法标准》GB/T 50123—2019。

7.3.3 Unconfined Compressive Strength Test

The unconfined compressive strength test is similar to the undrained test with $\sigma_3=0$ conducted in a triaxial apparatus. During the test, the cylindrical sample is placed in the unconfined compressive testing apparatus and vertical pressure is applied without any lateral pressure until the sample is sheared and destroyed. It is a special case of triaxial compression test. The maximum axial pressure q_u that the sample can withstand during shear failure is called the unconfined compressive strength. In the test, σ_3 remained constant at zero, and the principal stress was equal to the axial pressure. According to the test results, only one ultimate stress circle can be made ($\sigma_1=q_u$, $\sigma_3=0$) . For saturated cohesive soil, according to the results of the triaxial unconsolidated undrained test, its failure envelope is approximately a horizontal line, that is, $\varphi_u=0$. In this way, for the sole purpose of determining the undrained shear strength of saturated cohesive soil, a relatively simple unconfined compression tester can be used instead of a triaxial compression tester. At this point, assuming $\varphi_u=0$, the horizontal tangent of the ultimate stress circle obtained from the unconfined compressive strength test is the failure envelope (Figure 7.9) . Therefore,

7.3.3 无侧限抗压强度试验

无侧限抗压强度试验类似于在三轴仪中进行的 $\sigma_3=0$ 的不排水试验。在试验过程中，将圆柱形试样放置在无侧限抗压试验装置中，在没有任何侧向压力的情况下施加竖向压力，直到试样剪切破坏，这是三轴压缩试验的一个特例。试样在剪切破坏过程中能够承受的最大轴向压力 q_u 称为无侧限抗压强度。在试验中，σ_3恒定为零，大主应力等于轴向压力。根据试验结果，只能作一个极限应力圆（$\sigma_1=q_u$，$\sigma_3=0$）。对于饱和黏性土，根据三轴不固结不排水试验的结果，其破坏包络线近似为水平线，即 $\varphi_u=0$。这样，为了确定饱和黏性土的不排水抗剪强度，可以使用相对简单的无侧限抗压试验仪代替三轴压缩仪。此时，假设 $\varphi_u=0$，从无侧限抗压强度试验中获得的极限应力圆的水平切线是破坏包络线（图 7.9），其剪切强度可表示为式（7.18），

$$\tau_f = c_u = q_u/2 \tag{7.18}$$

where c_u is the undrained shear strength of soil (kPa); q_u is the unconfined compressive strength (kPa).

式中，c_u是土的不排水抗剪强度(kPa)；q_u为无侧限抗压强度(kPa)。

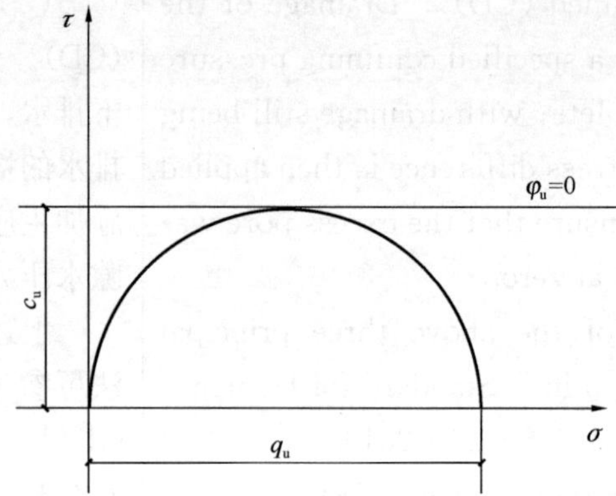

Figure 7.9 Unconfined compressive strength test results

Problems

7.1 A consolidated-drained triaxial test was conducted on a normally consolidated clay.

The results are as follows:

$\sigma_3 = 276\text{kN/m}^2$

$(\Delta\sigma_d)_f = 276\text{kN/m}^2$

Determine

(1) Angle of friction, φ'.

(2) Angle α that the failure plane makes with the major principal plane.

7.2 Refer to Problem 7.1.

(1) Find the normal stress σ' and the shear stress τ_f on the failure plane.

(2) Determine the effective normal stress on the plane of maximum shear stress.

7.3 The equation of the effective stress failure envelope for normally consolidated clayey soil is $\tau_f = \sigma' \tan 30°$. A drained triaxial test was conducted with the same soil at a chamber-confining pressure of 69kN/m^2. Calculate the deviator stress at failure.

7.4 The results of two drained triaxial tests on a saturated clay follow:

Specimen Ⅰ:

$$\sigma_3 = 70\text{kN/m}^2$$
$$(\Delta\sigma_d)_f = 130\text{kN/m}^2$$

Specimen Ⅱ:

$$\sigma_3 = 160\text{kN/m}^2$$
$$(\Delta\sigma_d)_f = 223.5\text{kN/m}^2$$

Determine the shear strength parameters.

7.5 A sample of saturated sand was consolidated under an all-around pressure of 105 kN/m^2. The axial stress was then increased and drainage was prevented. The specimen failed when the axial deviator stress reached 70 kN/m^2. The pore water pressure at failure was 50 kN/m^2. Determine:

(1) Consolidated-undrained angle of shearing resistance, φ.

(2) Drained friction angle, φ'.

Chapter 8 Earth Pressure

8.1 Introduction

Earth pressure refers to the lateral pressure exerted by the self-weight of the soil on the retaining structure, and is the main load acting on the retaining structure. The correct design and construction of these structures require determining the magnitude, direction, and point of application of earth pressure. The magnitude and distribution of lateral earth pressure depend on many factors, such as the possible displacement of the retaining structure under soil pressure, the properties of the soil being retained, and the form and stiffness of the retaining structure.

Earth retaining structures are widely used in building construction, bridges, roads, and water conservancy projects, such as retaining walls, basement side walls, bridge piers, etc. that support the surrounding soil of buildings or serve as slope support structures in mountainous areas (Figure 8.1). This chapter is devoted to the study of the various earth pressure theories and the calculation method of lateral earth pressure.

第 8 章 土压力

8.1 概述

土压力是指土因自重对挡土结构物产生的侧向压力，是作用于挡土结构物上的主要荷载。挡土结构的设计和施工需要确定土压力的大小、方向和作用点。侧向土压力的大小和分布取决于许多因素，如土压力作用下挡土结构物可能的位移、被挡土的性质以及挡土结构物的形式、刚度等。

挡土结构在房屋建筑、桥梁、道路以及水利等工程中有广泛的应用，例如，支撑建筑物周围填土或作为山区边坡支挡结构的挡土墙、地下室侧墙、桥台等（图 8.1）。本章将介绍各种土压力理论以及土压力计算方法。

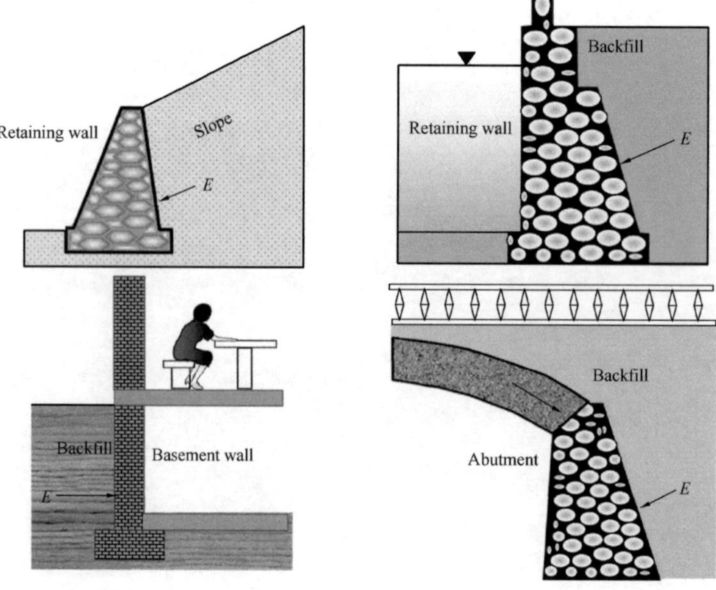

Figure 8.1 Application of retaining structures in engineering

8.2 At-Rest, Active, and Passive Pressures

Three possible cases may arise concerning the retaining wall shown as in Figure 8.2. A mass of soil is bounded by a frictionless wall of height AB. A soil element located at a depth z is subjected to a vertical pressure, σ_z, and a horizontal pressure, σ_x. There are no shear stresses on the vertical and horizontal planes of the soil element. Let us define the ratio of σ_x to σ_z as a nondimensional quantity K, or

8.2 静止、主动和被动土压力的概念

挡土墙可能出现的三种情况如图 8.2 所示。一土体以高度为 AB 的无摩擦墙为边界，位于深度 z 处的土体单元承受竖向压力 σ_z 和水平压力 σ_x。土体单元的竖直面和水平面上没有剪应力。将 σ_x 与 σ_z 的比值定义为无量纲量数 K，由式（8.1）表示。

$$K=\frac{\sigma_x}{\sigma_z} \tag{8.1}$$

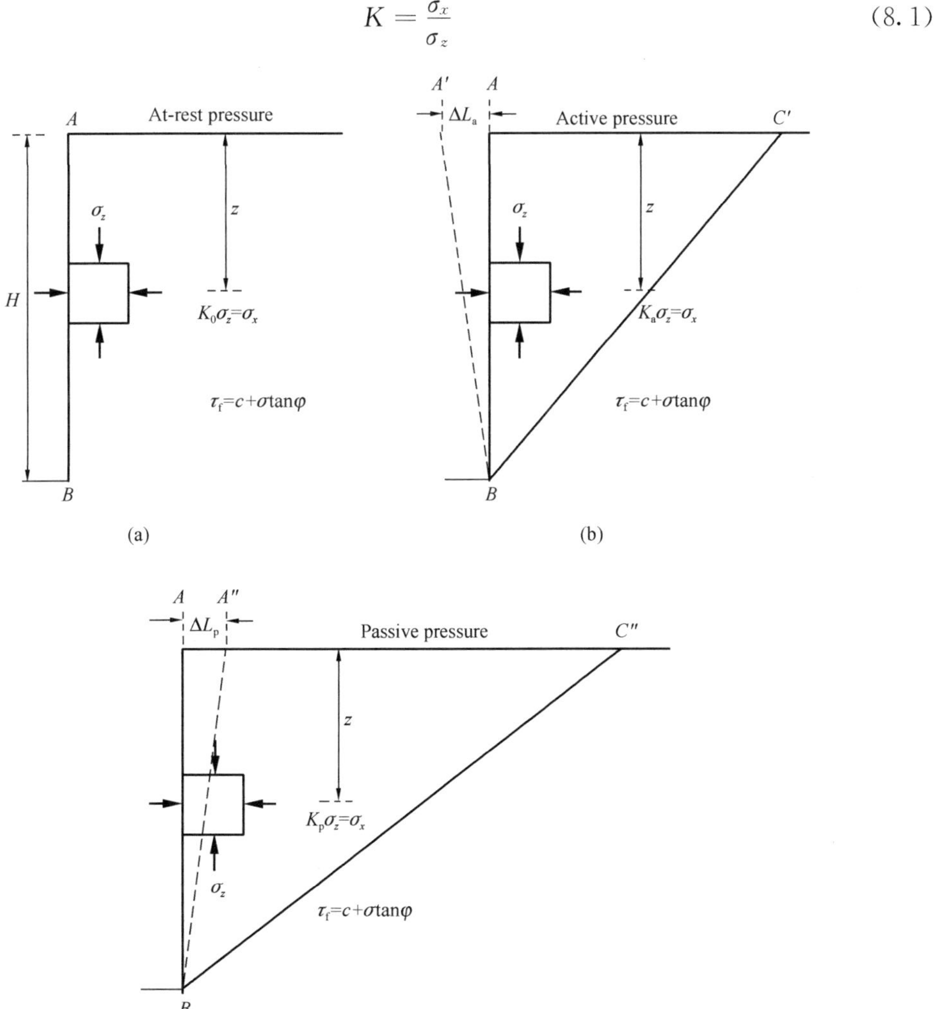

Figure 8.2 Definition of at-rest, active, and passive pressures (*Note*: Wall AB is frictionless)

(1) If the wall AB is static (Figure 8.2 a), that is, if it does not move either to the right or to the left of its initial position-the soil mass will be in a state of static equilibrium. In that case, σ_x is referred to as the at-rest earth pressure, or

（1）如果墙 AB 是静态的（图 8.2a），也就是说它不向初始位置的右侧或左侧移动，则土体将处于静态平衡状态。在这种情况下，σ_x 被称为静止土压力，由式（8.2）表示。

$$K = K_0 = \frac{\sigma_x}{\sigma_z} \tag{8.2}$$

where K_0 is at-rest earth pressure coefficient.

式中，K_0 为静止土压力系数。

(2) If the frictionless wall rotates sufficiently about its bottom to a position of $A'B$ (Figure 8.2b), then a triangular soil mass ABC' adjacent to the wall will reach a state of plastic equilibrium and will fail sliding down the plane BC'. At this time, the horizontal stress, $\sigma_x = \sigma_a$, will be referred to as active pressure. Now,

（2）如果无摩擦墙体围绕其底部充分旋转到 $A'B$ 的位置（图 8.2b），则与墙相邻的三角形土体 ABC' 将达到塑性平衡状态，并将无法沿 BC' 平面滑动。此时，水平应力 $\sigma_x = \sigma_a$ 将被称为主动土压力。由式（8.3）表示。

$$K = K_a = \frac{\sigma_x}{\sigma_z} = \frac{\sigma_a}{\sigma_z} \tag{8.3}$$

where K_a is active earth pressure coefficient.

式中，K_a 为主动土压力系数。

(3) If the frictionless wall rotates sufficiently about its bottom to a position $A''B$ (Figure 8.2c), then a triangular soil mass ABC'' will reach a state of *plastic equilibrium* and will fail sliding upward along the plane BC''. The horizontal stress at this time will be $\sigma_x = \sigma_p$, the so-called passive pressure. In this case,

（3）如果无摩擦墙体围绕其底部充分旋转到位置 $A''B$（图 8.2c），则三角形土体 ABC'' 将达到塑性平衡状态，并将无法沿平面 BC'' 向上滑动。此时的水平应力为 $\sigma_x = \sigma_p$，即被动土压力。在这种情况下，有式（8.4）。

$$K = K_p = \frac{\sigma_x}{\sigma_z} = \frac{\sigma_p}{\sigma_z} \tag{8.4}$$

where K_p is passive earth pressure coefficient

式中，K_p 为被动土压力系数。

Figure 8.3 shows the nature of variation of lateral earth pressure with the wall tilt. From Figure 8.3, it can be seen that under the same conditions, the displacement caused by passive earth pressure is much greater than that caused by active earth pressure. Therefore, the active earth pressure E_a acting on the retaining wall is less than the at-rest earth pressure E_0, and the at-rest earth pressure E_0 is less than the passive earth pressure E_p, that is, $E_a < E_0 < E_p$.

图 8.3 显示了侧向土压力随墙倾斜度变化的性质。从图 8.3 中可以看出，在相同条件下，被动土压力引起的位移远大于主动土压力造成的位移。因此，作用在挡土墙上的主动土压力 E_a 小于静止土压力 E_0，静止土压力 E_0 小于被动土压力 E_p，即 $E_a < E_0 < E_p$。

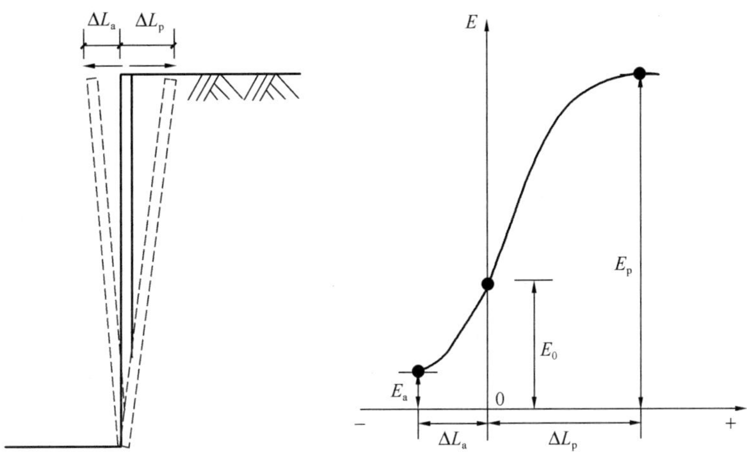

Figure 8.3 Variation of the magnitude of lateral earth pressure with wall tilt

8.3 At-Rest Earth Pressure

In order to define the at-rest earth pressure, if a unit cell is taken at any depth z below the surface of the backfill soil on the wall, and the vertical self-weight stress γz of the soil acts on it (Figure 8.4), the earth pressure strength at that point can be expressed as,

8.3 静止土压力

为了定义静止土压力，在墙体上填土面以下任意深度 z 处取一个单元，土的竖向自重应力 γz 作用于其上，如图 8.4 所示，则该点的土压力强度可表示为式 (8.5)。

$$\sigma_z = K_0 \gamma z \tag{8.5}$$

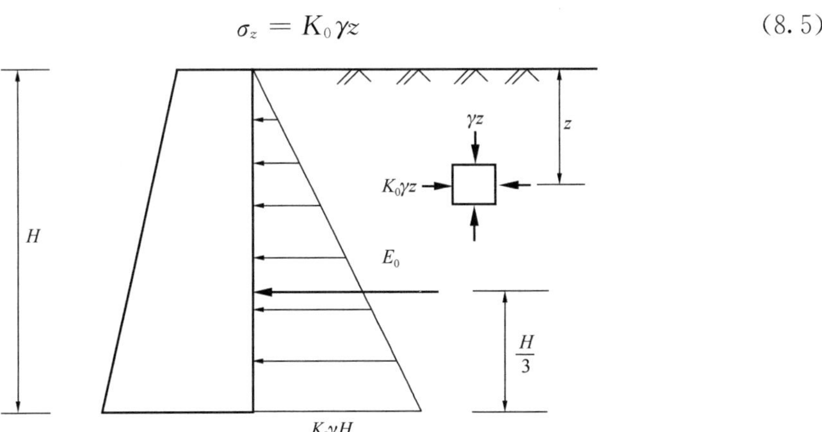

Figure 8.4 Distribution of earth pressure at-rest

where σ_z is the at-rest earth pressure strength (kPa); K_0 is at-rest earth pressure coefficient. For coarse-grained soils, the at-rest earth pressure coefficient can be estimated by using the empirical relationship

式中，σ_z 为静止土压力强度 (kPa)；K_0 为静止土压力系数。对于粗粒土，静止土压力系数可用经验关系式 (8.6) 估计，

$$K_0 = 1 - \sin\varphi \tag{8.6}$$

where φ is friction angle.

According to Eq. 8.5, the at-rest earth pressure is distributed in a triangular shape along the height of the wall. As shown in Figure 8.4, taking a unit length of retaining wall, the at-rest earth pressure acting on the wall is

式中，φ 为摩擦角。

由式（8.5）可知，静止土压力沿墙高呈三角形分布。如图 8.4 所示，取单位长度挡土墙，其作用于墙体的静止土压力为式（8.7）。

$$E_0 = (1/2)\ \gamma H^2 K_0 \tag{8.7}$$

Where E_0 is the at-rest earth pressure (kN/m), which point of action is at a distance of $H/3$ from the bottom of the wall; H is the height of retaining wall.

式中，E_0 为静止土压力（kN/m），作用点距墙底 $H/3$ 处，H 为挡土墙高度。

8.4 Rankine's Theory of Earth Pressure

In 1857, Rankine investigated the stress conditions in soil at a state of plastic equilibrium and proposed the theory of earth pressure, known as the Rankine's theory of earth pressure. In the following sections, we deal with Rankine's theory of earth pressure.

Let us examine a half space with a horizontal surface bounded by a frictionless wall, where the soil extends downwards and horizontally to infinity as shown in Figure 8.5 (a). When the entire soil is in a static state, all points are in an elastic equilibrium state. Taking a unit cell at a depth of z from the ground surface, assuming the unit weight of soil is γ, the normal stress on the horizontal section of the unit cell is equal to the self-weight stress of the soil at that point, that is $\sigma_z = \gamma z$. While the horizontal normal stress on the vertical section is equivalent to the earth pressure at-rest strength, which $\sigma_x = K_0 \gamma z$. Due to the fact that each vertical plane in the half space soil is a symmetrical plane, the shear stress on the vertical and horizontal sections of the soil element is equal to zero, and the normal stresses σ_z and σ_x on the corresponding sections are principal stresses. This point is in an elastic equilibrium state and can be represented by the Mohr's circle Ⅰ as shown in Figure 8.5(b). At this point, the Mohr's circle is located below the shear strength envelope.

8.4 朗肯土压力理论

1857 年，Rankine 研究了土体在塑性平衡状态下的应力状态，提出了土压力理论，称为朗肯土压力理论。在下面几节中，讨论朗肯土压力理论。

假设有一个水平面以光滑墙体为边界的半空间，其中土体向下和水平延伸到无限远，如图 8.5（a）所示。当整个土体处于静力平衡状态时，各点均处于弹性平衡状态。取距地面深度为 z 的单元，假设土的重度为 γ，则单元水平截面上的正应力等于该点土的自重应力，即 $\sigma_z = \gamma z$。竖向截面上的水平正应力相当于静止土压力强度，$\sigma_x = K_0 \gamma z$。由于半空间土中各竖向平面均为对称平面，因此土单元在竖向和水平截面上的剪应力均为零，对应截面上的正应力 σ_z 和 σ_x 为主应力。该点处于弹性平衡状态，可用莫尔圆Ⅰ表示，如图 8.5（b）所示。此时，莫尔圆位于抗剪强度包络线的下方。

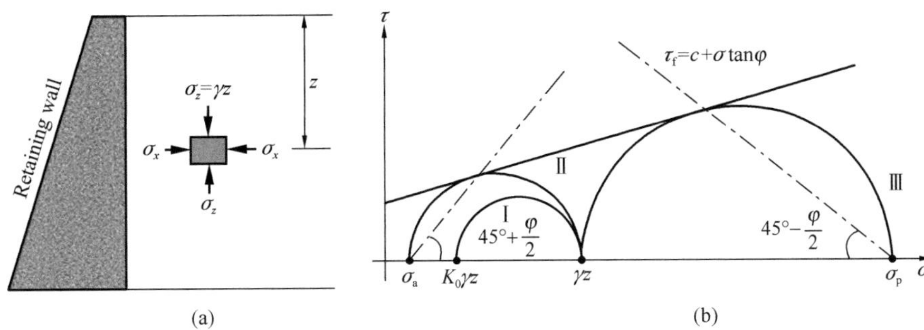

Figure 8. 5 Rankine′s earth pressure

If the retaining wall is displaced away from the soil mass and causes the entire soil mass to extend uniformly in the horizontal direction, the normal stress σ_x on the vertical section of the soil element gradually decreases, and the normal stress σ_z on the horizontal section remains constant, thus reaching the ultimate equilibrium condition. At this point, the Mohr′s circle is tangent to the shear strength envelope with σ_x and σ_z being the minor and major principal stresses, respectively, as shown circle Ⅱ in Figure 8. 5 (b), which is called the active Rankine state. On the contrary, if the displacement of the retaining wall towards the soil mass causes uniform compression of the entire soil mass in the horizontal direction, then σ_x increases continuously while σ_z remains constant until the limit equilibrium condition is met, which is called the passive Rankine state, as shown circle Ⅲ in Figure 8. 5 (b) .

如果挡土墙与向远离土体方向位移，使整个土体在水平方向上均匀伸展，土单元竖直截面上的正应力 σ_x 将逐渐减小，水平截面上的正应力 σ_z 保持不变，达到最终平衡状态。此时，莫尔圆与抗剪强度包络线相切，σ_x 和 σ_z 为主应力。如图 8.5（b）中莫尔圆Ⅱ所示，称为主动朗肯状态。反之，如果挡土墙向土体位移，则整个土体在水平方向受到均匀压缩，σ_x 不断增大，σ_z 保持不变，直至达到极限平衡条件，称为被动朗肯状态，如图 8.5（b）中莫尔圆Ⅲ所示。

Noted that only retaining walls with back sliding, upright, and horizontal filling surfaces can meet the active or passive Rankine boundary conditions. Therefore, the Rankine′s earth pressure theory is applicable to such retaining walls with back sliding, upright, and horizontal filling surfaces.

需要注意的是，只有墙背光滑、直立和填土面水平的挡土墙才能满足主动或被动朗肯边界条件。因此，朗肯土压力理论适用于墙背光滑、直立、填土面水平的挡土墙。

8. 4. 1 Active Earth Pressure

8. 4. 1 主动土压力

For the retaining wall shown in Figure 8. 5 (a), when the retaining wall is far away from the soil, the vertical stress σ_z at any depth z of the backfill behind the wall remains constant, which is the major principal stress σ_1. The horizontal stress σ_x gradually decreases

对于图 8.5（a）所示的挡土墙，当挡土墙远离土体时，墙后填土在任意深度处的竖向应力 σ_z 保持不变，为大主应力 σ_1。水平应力 σ_x 逐渐减小，直至达到

until it reaches the active Rankine state, where σ_x is the small principal stress σ_3, which is the active earth pressure strength σ_a. Therefore, according to the formula for limit equilibrium conditions, it can be obtained that for non-cohesive soil,

主动朗肯状态，其中 σ_x 为小主应力σ_3，为主动土压强度σ_a。因此，根据极限平衡条件，对于无黏性土，可得式（8.8）。

$$\sigma_a = \gamma z K_a \tag{8.8}$$

for cohesive soil,

对于黏性土，可得式（8.9）。

$$\sigma_a = \gamma z K_a - 2c\sqrt{K_a} \tag{8.9}$$

$$K_a = \tan^2\left(45° - \frac{\varphi}{2}\right) \tag{8.10}$$

where σ_a is the active earth pressure (kPa); K_a is the coefficient of Rankine′s active earth pressure; γ, c and φ are respectively the unit weight, cohesion force and internal friction angle of backfill behind the wall (kN/m^3, kPa, °); z is the depth of the reference point away from the fill surfae.

式中，σ_a为主动土压力（kPa），K_a为朗肯主动土压力系数，由式（8.10）表示。γ、c、φ 分别为墙后填土重度（kN/m^3）、黏聚力（kPa）、内摩擦角（°），z 是参考点离填土面的深度。

According to Eq. (8.8), the active soil pressure of non-cohesive soil is proportional to z and distributed in a triangular shape along the wall height as shown in Figure 8.6.

由式（8.8）可知，无黏性土的主动土压力与 z 成正比，沿墙高呈三角形分布，如图 8.6 所示。

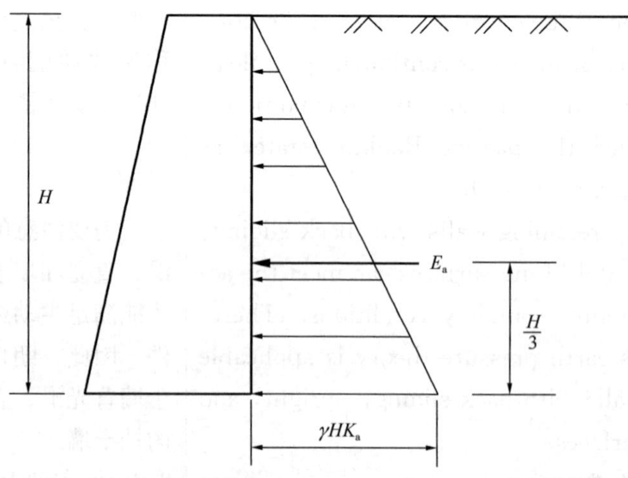

Figure 8.6 Active earth pressure distribution of non-cohesive soil

The total active earth force of non-cohesive soil per unit length of the wall can be expressed as

墙体单位长度无黏性土的总主动土力可表示为式（8.11）。

$$E_a = \frac{1}{2}\gamma H^2 K_a \tag{8.11}$$

where E_a is total active earth force of non-cohesive soil, whose line of action passes through the centroid of the triangle, that is acts at a distance of $H/3$ measured from the bottom of the wall.

式中，E_a 为无黏性土的总主动土力，其作用线经过三角形形心，即距离墙底 $H/3$ 处。

From Eq. (8.9), the active earth pressure of cohesive soil includes two parts, one is $\gamma z K_a$ caused by the self-weight of the soil, and the other is the negative lateral pressure $2c\sqrt{K_a}$ caused by the cohesion of the soil. The result of the superposition of these two parts of earth pressure is shown in Figure 8.7.

由式（8.9）可知，黏性土的主动土压力包括两部分，一部分是土自重引起的 $\gamma z K_a$，另一部分是土的黏性引起的负侧压力 $2c\sqrt{K_a}$。这两部分土压力的叠加结果如图 8.7 所示。

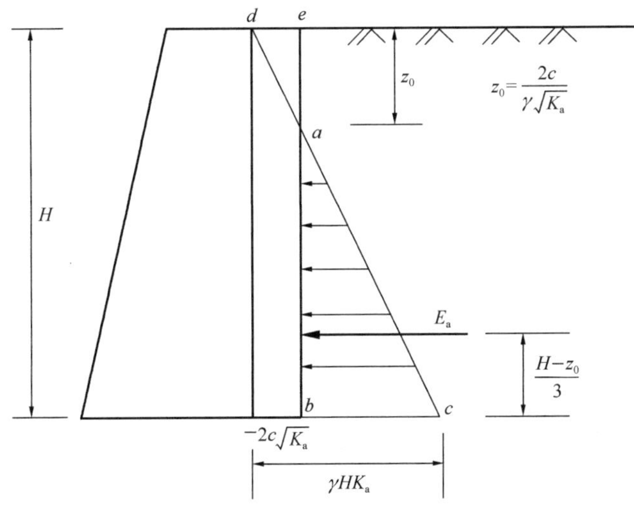

Figure 8.7 Active earth pressure distribution of cohesive soil

In this way, there is a point along the height of the wall where the earth pressure is 0, and the depth of this point is called the critical depth, which can be derived by Eq. (8.9) equal to 0, that is

这样，沿墙高方向存在一个土压力为 0 的点，该点的深度称为临界深度，可由式（8.9）等于 0 导出，导出式（8.12），

$$\sigma_a = \gamma z K_a - 2c\sqrt{K_a} = 0$$

$$z_0 = \frac{2c}{\gamma\sqrt{K_a}} \tag{8.12}$$

The total active earth force of cohesive soil per unit length of the wall is

单位长度墙体黏性土的总主动土力为式（8.13）。

$$E_a = \frac{1}{2}\gamma (H - z_0)^2 K_a \tag{8.13}$$

where E_a is total active earth force of cohesive soil, whose line of action passes through the centroid of the Δabc, that is acts at a distance of $(H-z_0)/3$ measured from the bottom of the wall.

式中，E_a 为黏性土的总主动土力，其作用线经过 Δabc 的形心，即作用于距离墙底 $(H-z_0)/3$ 处。

Example 8. 1

There is a retaining wall with a vertical, smooth back and a horizontal filling surface, with a height of 6 meters. The physical and mechanical properties of the backfill behind the wall are as follows:

$\gamma=20\text{kN/m}^3$, $c=10\text{kN/m}^3$, $\varphi=30°$. Calculate the active earth pressure and determine its point of action, and draw a distribution diagram of the active earth pressure.

例题 8. 1

有一个挡土墙，墙背竖直、光滑，填土面水平，高度为6m。墙后填土的物理力学性能如下：

$\gamma=20\text{kN/m}^3$，$c=10\text{kN/m}^3$，$\varphi=30°$。计算主动土压力并确定其作用点，并绘制主动土压力分布图。

Solution

解答

$$K_a=\tan^2\left(45°-\frac{30°}{2}\right)=\frac{1}{3},\ z_0=\frac{2\times 10}{20\sqrt{1/3}}=\sqrt{3}\text{m}$$

$$\sigma_{a0}=-2\times 10\times\sqrt{1/3}\approx 11.55\text{kPa}$$

$$\sigma_{a1}=20\times 6\times\frac{1}{3}-2\times 10\times\sqrt{1/3}\approx 28.45\text{kPa}$$

$$E_a=\frac{1}{2}(6-\sqrt{3})(20\times 6\times\frac{1}{3}-2\times 10\times\sqrt{1/3})$$

$$\approx 60.72\text{kN/m}$$

The distribution diagram of the active earth pressure is shown in Figure 8. 8.

主动土压力分布如图 8. 8 所示。

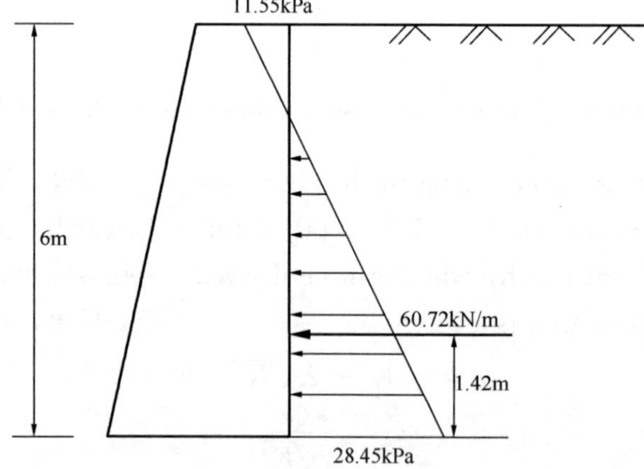

Figure 8. 8 Distribution diagram of the active earth pressure

8. 4. 2 Passive Earth Pressure

When the retaining wall as shown in Figure 8. 5(a) moves towards the soil and squeezes the backfill behind the wall, the vertical stress $\sigma_z=\gamma z$ at any point in the soil remains constant, which is the minor principal stress σ_3, while the horizontal stress σ_x gradually

8. 4. 2 被动土压力

如图 8. 5(a) 所示挡土墙向土体移动并挤压墙后填土时，土体任意点处的竖向应力 $\sigma_z=\gamma z$ 保持不变，为小主应力 σ_3，而水平应力 σ_x 逐渐增大，直至达

increases until it reaches the passive Rankine's state. At this point, the horizontal stress reaches its maximum value, which is the major principal stress σ_1, which is the passive earth pressure. Therefore, according to the formula for limit equilibrium conditions, it can be derived that:

到被动朗肯状态。此时水平应力达到最大值，即大主应力 σ_1，为被动土压力。因此，根据极限平衡条件：

for non-cohesive soil,

对无黏性土，可推导出式 (8.14)。

$$\sigma_p = \gamma z K_p \tag{8.14}$$

for cohesive soil,

对于黏性土，可得式 (8.15)。

$$\sigma_p = \gamma z K_p + 2c\sqrt{K_p} \tag{8.15}$$

$$K_p = \tan^2\left(45° + \frac{\varphi}{2}\right) \tag{8.16}$$

where σ_p is the passive earth pressure (kPa); K_p is the coefficient of Rankine's passive earth pressure.

式中，σ_p为被动土压力 (kPa)；K_p为朗肯被动土压力系数，由式 (8.16) 表示。

On the basis of Eq. (8.14) and Eq. (8.15), the passive earth pressure of non-cohesive soil is distributed in a triangular shape along the wall height (Figure 8.9a), while the passive earth pressure of cohesive soil is composed of two parts: one is generated by its own weight and the other is generated by cohesive force. Therefore, its passive earth pressure is distributed in a trapezoidal shape along the wall height (Figure 8.9b), in which x is the distance from the centroid of a trapezoid to the bottom of the wall.

由式 (8.14) 和式 (8.15) 可知，无黏性土的被动土压力沿墙高呈三角形分布 (图 8.9a)，而黏性土的被动土压力由自重产生的土压力和黏聚力产生的土压力两部分组成。因此其被动土压力沿墙高呈梯形分布 (图 8.9b)，其中 x 为梯形形心到墙底的距离。

The total passive earth force of soil per unit length of the wall is

墙单位长度土体的总被动土压力，可用式 (8.17) 和式 (8.18) 表示。

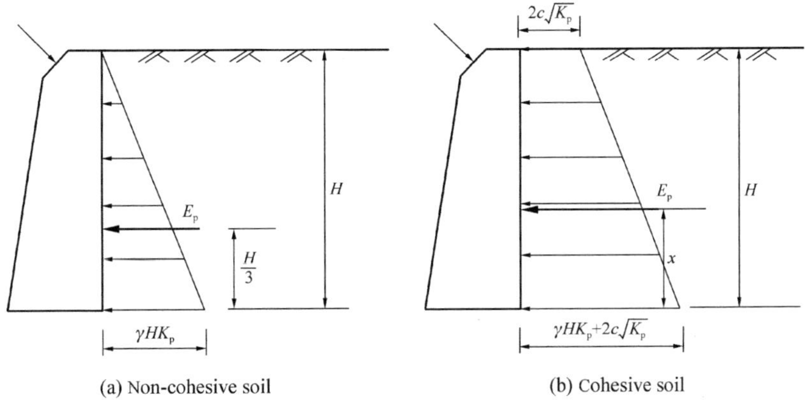

Figure 8.9 Distribution diagram of the passive earth pressure

$$E_p = \frac{1}{2}\gamma H^2 K_p \text{ for non-cohesive soil} \tag{8.17}$$

$$E_p = \frac{1}{2}\gamma H^2 K_p + 2cH\sqrt{K_p} \text{ for cohesive soil} \tag{8.18}$$

where E_p is total passive earth force of soil, whose line of action passes through the centroid of the distribution diagram of the passive earth pressure.

式中，E_p 为总被动土压力，其作用线经过被动土压力分布图的形心。

Example 8. 2

例题 8. 2

There is a retaining wall with a vertical, smooth back and a horizontal filling surface, with a height of 6 meters. The physical and mechanical properties of the backfill behind the wall are as follows:

γ=20kN/m^3, c=10kN/m^3, φ=30°. Calculate the passive earth pressure and determine its point of action, and draw a distribution diagram of the passive earth pressure.

有一个挡土墙，墙背竖直、光滑，填土面水平，高度为 6m。墙后填土的物理力学性能如下：

γ = 20kN/m^3，c = 10kN/m^3，φ=30°. 计算被动土压力并确定其作用点，并绘制被动土压力分布图。

Solution

解答

$$K_p = \tan^2\left(45° + \frac{30°}{2}\right) = 3, \quad \sigma_{p0} = 2 \times 10 \times \sqrt{3} \approx 34.64\text{kPa}$$

$$\sigma_{a1} = 20 \times 6 \times 3 + 2 \times 10 \times \sqrt{3} \approx 394.64\text{kPa}$$

$$E_p = \frac{1}{2} \times 20 \times 6^2 \times 3 + 2 \times 10 \times 6 \times \sqrt{3} \approx 1287.85\text{kN/m}$$

$$x = \frac{\left(\frac{1}{2} \times 20 \times 6^2 \times 3 \times \frac{6}{3} + 2 \times 10 \times 6 \times \sqrt{3} \times \frac{6}{2}\right)}{1287.85} = 2.16\text{m}$$

The distribution diagram of the passive earth pressure is shown in Figure 8. 10.

被动土压力分布如图 8. 10 所示。

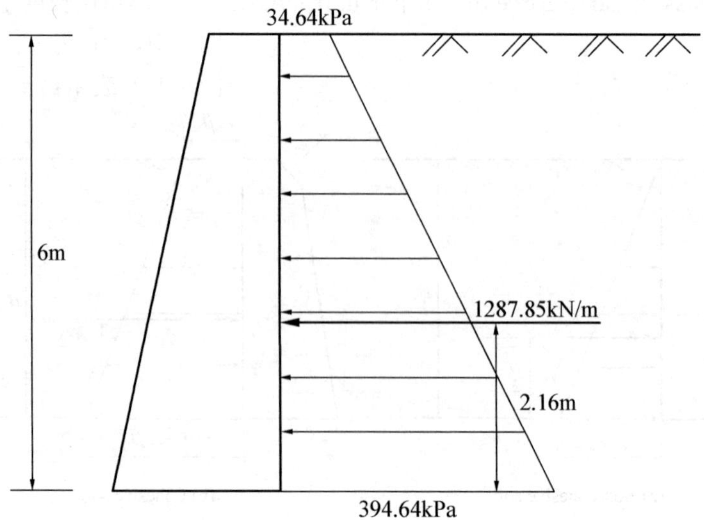

Figure 8. 10 Distribution diagram of the passive earth pressure

8.4.3 The Earth Pressure When There is an Overload Act on the Filling Surface

When there is a continuous uniformly distributed load q acting on the backfill surface behind the retaining wall (Figure 8.11), the uniformly distributed load is usually converted into equivalent soil weight, that is, the imaginary soil weight is used instead of the uniformly distributed load. The equivalent soil layer thickness is

8.4.3 填土面有超载作用时的土压力

当挡土墙后填土面承受连续均布荷载 q 时（图 8.11），通常将均布荷载转化为等效土重，即用假想土重代替均布荷载。等效土层厚度可表达为式（8.19）。

$$h = q/\gamma \tag{8.19}$$

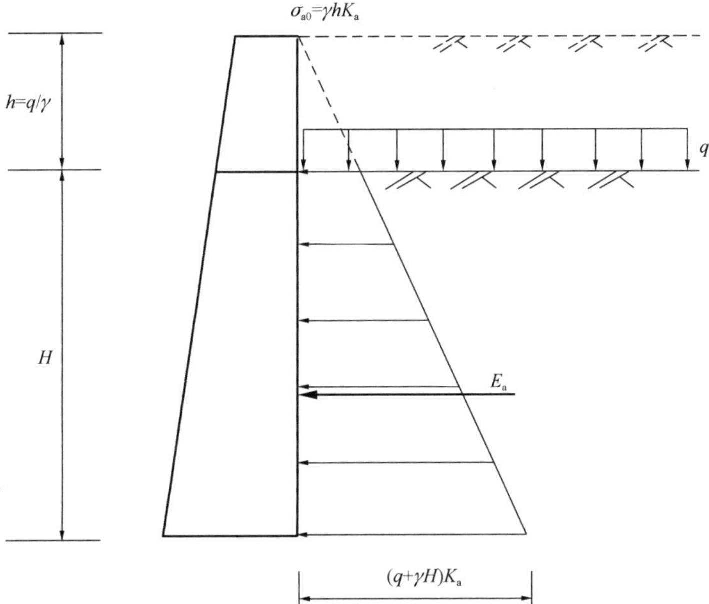

Figure 8.11 Distribution of active earth pressure when there is a continuous uniformly distributed load q acting on the backfill surface

If the backfill behind the wall is non-cohesive soil, the active earth pressure at the top of the wall is

若墙后填土为无黏性土，则墙顶主动土压力为式（8.20），

$$\sigma_{a0} = \gamma h K_a \tag{8.20}$$

The active earth pressure at the bottom of the wall is

墙底主动土压力为式（8.21）。

$$\sigma_{aH} = \gamma(h + H)K_a = (q + \gamma H)K_a \tag{8.21}$$

8.4.4 The Earth Pressure of Nonhomogeneous Soil Behind the Wall

When the backfill behind the wall is layered (Figure 8.12), the earth pressure of the first layer is calculated as homogeneous soil. When calculating the earth pressure of the second layer, first convert the first layer of soil into an equivalent soil layer with the same density as the second layer of soil, that is, its equivalent soil layer thickness is $h'_1 = h_1\gamma_1/\gamma_2$, and then use $(h'_1 + h_2)$ as the wall height to calculate the earth pressure based on homogeneous soil. Note that the earth pressure coefficients of each layer of soil are different.

8.4.4 墙后非均质土土压力

当墙后填土分层时（图 8.12），第一层土压力按均质土计算。计算第二层土压力时，首先将第一层土转换为与第二层土密度相同的等效土层，即其等效土层厚度为 $h'_1 = h_1\gamma_1/\gamma_2$，然后以 $(h'_1 + h_2)$ 作为墙高，按均质土计算土压力。注意每一层土的土压力系数是不同的。

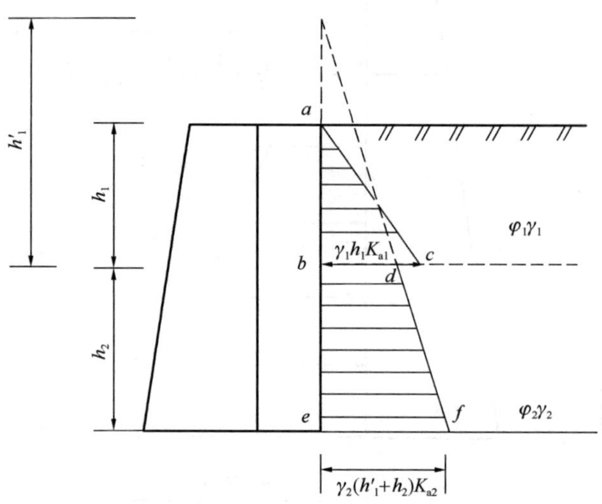

Figure 8.12 Distribution of active earth pressure when the backfills behind the wall is layered

When there is groundwater in the backfill behind the wall, there are two methods for calculating soil pressure. One method is soil and water separation calculation for non-cohesive soil, and the other method is soil and water sum calculation cohesive soil. As shown in Figure 8.13, the groundwater table is located at a depth of h_1 below the ground surface. The distribution of earth pressure calculated by the two methods is shown in Figure 8.13(a) and Figure 8.13(b) respectively.

当墙后填土中存在地下水时，土压力计算方法有两种。一种方法是对无黏性土进行水土分离计算，另一种方法是对黏性土采用水土总和法计算。如图 8.13 所示，地下水位位于地表以下 h_1 深度时，两种方法计算的土压力分布分别如图 8.13(a) 和图 8.13(b) 所示。

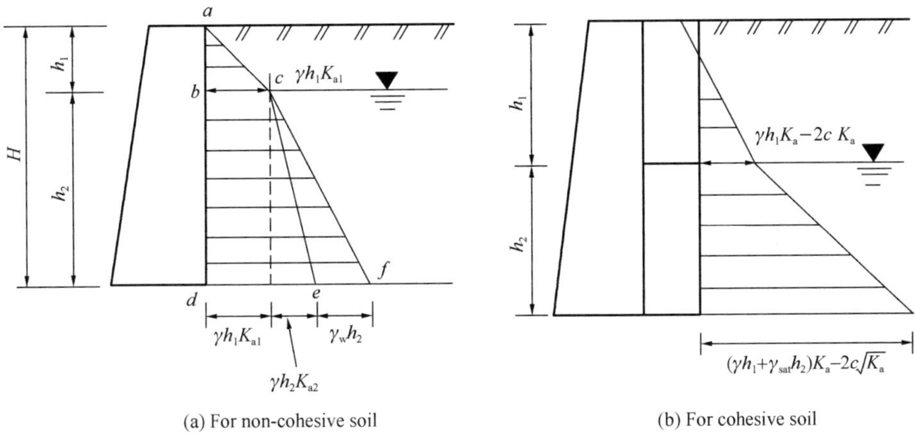

Figure 8. 13 Distribution of active earth pressure when there is groundwater in the backfill behind the wall

Example 8. 3

For the retaining wall shown in Figure 8. 14, determine the force per unit length of the wall for Rankine's active state. Also find the location of the resultant.

例题 8. 3

对于图 8. 14 所示的挡土墙，在朗肯主动状态下确定单位长度墙体的力，并确定合力作用点的位置。

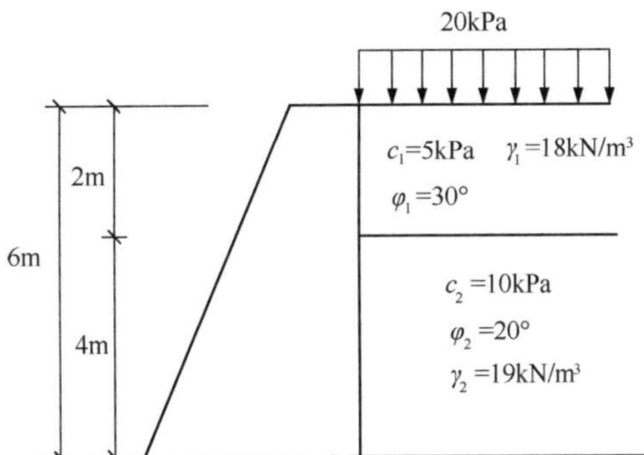

Figure 8. 14 The retaining wall

Solution

For the upper layer of the soil, Rankine's active earth-pressure coefficient is

$$K_{a1} = \tan^2\left(45° - \frac{\varphi}{2}\right) = \tan^2\left(45° - \frac{30°}{2}\right) \approx 0.333$$

For the lower layer,

$$K_{a2} = \tan^2\left(45° - \frac{\varphi}{2}\right) = \tan^2\left(45° - \frac{20°}{2}\right) \approx 0.490$$

At $z=0$,

$$\sigma_{a0} = qK_{a1} - 2c_1\sqrt{K_{a1}} = 20 \times 0.333 - 2 \times 5 \times \sqrt{0.333} = 0.889\text{kPa}$$

At $z=2.0$m (just inside the bottom of the upper layer),

$$\sigma_{a1}=(q+\gamma_1 h_1)K_{a1}-2c_1\sqrt{K_{a1}}=(20+18\times 2)\times 0.333-2\times 5\times\sqrt{0.333}=12.877\text{kPa}$$

At $z=2.0$m (in the lower layer),

$$\sigma'_{a1}=(q+\gamma_1 h_1)K_{a2}-2c_2\sqrt{K_{a2}}=(20+18\times 2)\times 0.49-2\times 10\times\sqrt{0.49}=13.44\text{kPa}$$

At $z=6.0$m,

$$\sigma_{a2}=(q+\gamma_1 h_1+\gamma_2 h_2)K_{a2}-2c_2\sqrt{K_{a2}}=(20+18\times 2+19\times 4)\times 0.49-2\times 10\times\sqrt{0.49}=50.68\text{kPa}$$

Thus, the total earth force is

$$P_a=\frac{1}{2}(\sigma_{a0}+\sigma_{a1})h_1+\frac{1}{2}(\sigma'_{a1}+\sigma_{a2})h_2=\frac{1}{2}(0.889+12.877)\times 2+\frac{1}{2}(13.44+50.68)\times 4=142.016\text{kN/m}$$

The location of the resultant can be found by taking the moment about the bottom of the wall:

$$\bar{z}=\frac{\left[\frac{1}{2}\times(0.889+12.877)\times 2\times\frac{6}{3}+\frac{1}{2}\times(13.44+50.68)\times 4\times\frac{6}{2}\right]}{142.016}=2.903$$

8.5 Coulomb′s Earth Pressure Theory

In 1776, Coulomb presented a theory for active and passive earth pressures against retaining walls. In this theory, Coulomb assumed that the failure surface is a plane. The wall friction was taken into consideration. The following sections discuss the general principles of the derivation of Coulomb′s earth pressure theory for a cohesionless backfill (shear strength defined by the equation $\tau_f=\sigma\tan\varphi$).

8.5.1 Coulomb′s Active Pressure

Let AB (Figure 8.15a) be the back face of a retaining wall supporting a granular soil, the surface of which is constantly sloping at an angle α with the horizontal. BC is a trial failure surface. In the stability consideration of the probable failure wedge ABC, the following forces are involved (per unit length of the wall): W is the weight of the soil wedge. R is the resultant of the shear and normal forces on the surface of failure. This is inclined at an angle of φ to the normal

8.5 库仑土压力理论

1776 年，库仑提出了挡土墙的主动和被动土压力理论。在该理论中，库仑假设破坏面是一个平面。考虑了墙壁摩擦力。下面几节讨论库仑土压力理论推导的一般原理，该理论适用于无黏性土（抗剪强度由方程 $\tau_f=\sigma\tan\varphi$ 定义）。

8.5.1 库仑主动压力

设 AB（图 8.15a）为支撑粗粒土的挡土墙墙背，挡土墙的表面与水平面呈倾斜状态（倾角为 α）。BC 为假设的破坏面。在考虑可能破坏的楔体 ABC 的稳定性时，涉及以下作用力（每单位墙长）：W 为楔体的自重。R 是破坏面的剪切力和法向力的合力。它与平面 BC 的法线之间的

drawn to the plane BC. E_a is the active force per unit length of the wall. The direction of E_a is inclined at an angle δ to the normal drawn to the face of the wall that supports the soil. δ is the angle of friction between the soil and the wall.

夹角为 φ。E_a是单位长度墙的主动土压力。E_a的方向与支撑土体的墙壁表面的法线呈 δ 角。δ 是土与墙之间的摩擦角。

The force triangle for the wedge is shown in Figure 8.15(b). From the law of sines, we have

楔形体的受力三角形如图 8.15(b) 所示。根据正弦定理，式 (8.22)

$$\frac{W}{\sin(90+\theta+\delta-\beta+\varphi)}=\frac{E_a}{\sin(\beta-\varphi)} \tag{8.22}$$

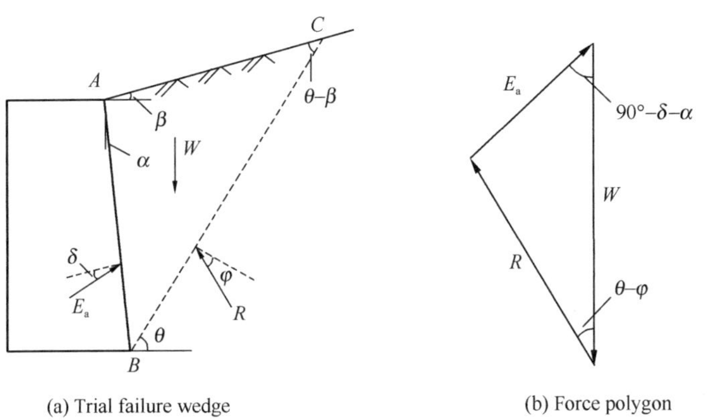

(a) Trial failure wedge

(b) Force polygon

Figure 8.15 Coulomb's active pressure

or

或式 (8.23) 成立，

$$E_a=\frac{\sin(\beta-\varphi)}{\sin(90+\theta+\delta-\beta+\varphi)}W \tag{8.23}$$

The preceding equation can be written in the form

式(8.23)可写为式(8.24)。

$$E_a=\frac{1}{2}\gamma H^2\left[\frac{\cos(\theta-\beta)\cos(\theta-\alpha)\sin(\beta-\varphi)}{\cos^2\theta\sin(\beta-\alpha)\sin(90+\theta+\delta-\beta+\varphi)}\right] \tag{8.24}$$

where γ is unit weight of the backfill. The values of γ, H, θ, α, φ, and δ are constants, and β is the only variable. To determine the critical value of β for maximum E_a, we have

式中，γ 为填土重度。γ，H，θ，α，φ 和 δ 的值是常数，β 是唯一的变量。为确定最大值 E_a 对应的 β 临界值，令

$$\frac{dE_a}{d\beta}=0 \tag{8.25}$$

After solving Eq. (8.25), when the relationship of β is substituted into Eq. (8.24), we obtain Coulomb's active earth pressure as

求解式 (8.25) 后，将 β 的关系式代入式 (8.24)，可得库仑主动土压力式 (8.26)。

$$E_a=\frac{1}{2}K_a\gamma H^2 \tag{8.26}$$

where K_a is Coulomb's active earth pressure coefficient and is given by

式中，K_a为库仑主动土压力系数，可表达为式（8.27）。

$$K_a=\frac{\cos^2(\varphi-\theta)}{\cos^2\theta\cos(\delta+\theta)\left[1+\sqrt{\frac{\sin(\delta+\varphi)\sin(\varphi-\alpha)}{\cos(\delta+\theta)\cos(\theta-\alpha)}}\right]^2} \tag{8.27}$$

Note that when $\alpha=0°$, $\theta=0°$, and $\delta=0°$, Coulomb's active earth pressure coefficient becomes equal to $\tan^2\left(45°-\frac{\varphi}{2}\right)$, which is the same as Rankine's earth pressure coefficient given earlier in this chapter.

注意，当$\alpha=0°$、$\theta=0°$、$\delta=0°$时，库仑主动土压力系数等于$\tan^2\left(45°-\frac{\varphi}{2}\right)$，与本章前面给出的朗肯主动土压力系数相同。

8.5.2 Coulomb's Passive Pressure

8.5.2 库仑被动压力

Figure 8.16(a) shows a retaining wall with a sloping cohesionless backfill similar to that considered in Figure 8.15 (a). The force polygon for equilibrium of the wedge ABC for the passive state is shown in Figure 8.16(b). E_p is the notation for the passive force. Other notations used are the same as those for the active case (Section 8.5.1). In a procedure similar to the one that we followed in the active case [Eq. (8.25)], we get

图 8.16(a) 为与图 8.15(a) 类似的填土为无黏性土的挡土墙。楔形体 ABC 在被动状态下的力的矢量多边形如图 8.16(b) 所示。E_p是被动土压力的符号。使用的其他符号与主动土压力情况相同（第 8.5.1 节）。与主动土压力情况类似［式（8.25)］，可得式（8.28）。

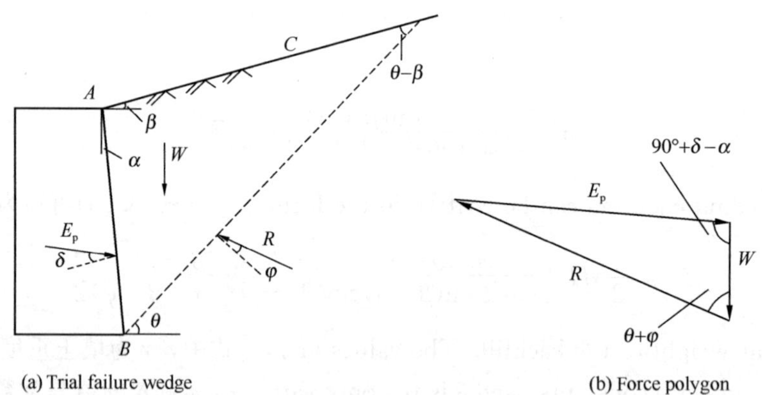

Figure 8.16 Coulomb's passive pressure

$$E_p=\frac{1}{2}K_p\gamma H^2 \tag{8.28}$$

where K_p = Coulomb's passive earth pressure coefficient, or

式中，K_p为库仑被动土压力系数，可表达为式（8.29）。

$$K_p=\frac{\cos^2(\varphi+\theta)}{\cos^2\theta\cos(\delta-\theta)\left[1-\sqrt{\frac{\sin(\varphi+\delta)\sin(\varphi+\alpha)}{\cos(\delta-\theta)\cos(\alpha-\theta)}}\right]^2} \tag{8.29}$$

For a frictionless wall with the vertical back face

对于墙面光滑，竖向墙背支

supporting granular soil backfill with a horizontal surface (that is, $\theta = 0°$, $\alpha = 0°$, and $\delta = 0°$), Eq. (8.30) yields

撑填土面水平（即 $\theta = 0°$，$\alpha = 0°$和 $\delta = 0°$）粗粒回填土的挡土墙，可得被动土压力系数式(8.30)。

$$K_p = \frac{1+\sin\varphi}{1-\sin\varphi} = \tan^2\left(45+\frac{\varphi}{2}\right) \tag{8.30}$$

This relationship is the same as that obtained for the passive earth pressure coefficient in Rankine's case, given by Eq. (8.16).

此式与式（8.16）给出的朗肯被动土压力系数相同。

Problems

8.1 For the retaining wall shown in Figure 8.17, determine the force per unit length of the wall for Rankine's active state. Also find the location of the resultant.

8.2 A frictionless retaining wall is shown in Figure 8.18. Determine:

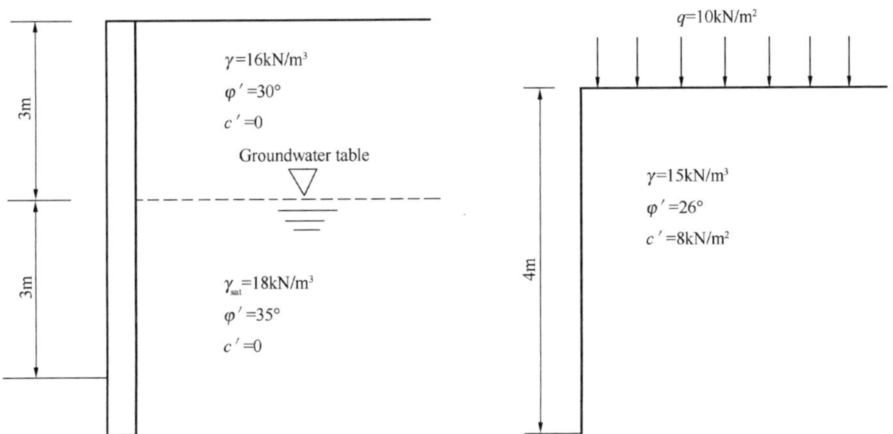

Figure 8.17 Figure of problem 8.1 Figure 8.18 Figure of problem 8.2

a. The active force E_a after the tensile crack occurs

b. The passive force E_p

8.3 Assume that the retaining wall shown in Figure 8.19 is frictionless. For each problem, determine the Rankine's passive force per unit length of the wall, the variation of active earth pressure with depth, and the location of the resultant.

Problem	H (m)	φ' (°)	γ (kN/m³)
8.3	3.35	32	18.4
8.4	4.87	38	15.8
8.5	7	30	16.6
8.6	12	27	20.5

8.4 A retaining wall is shown in Figure 8.20. For each problem, determine the Rankine's active force, E_a, per unit length of the wall and the location of the resultant.

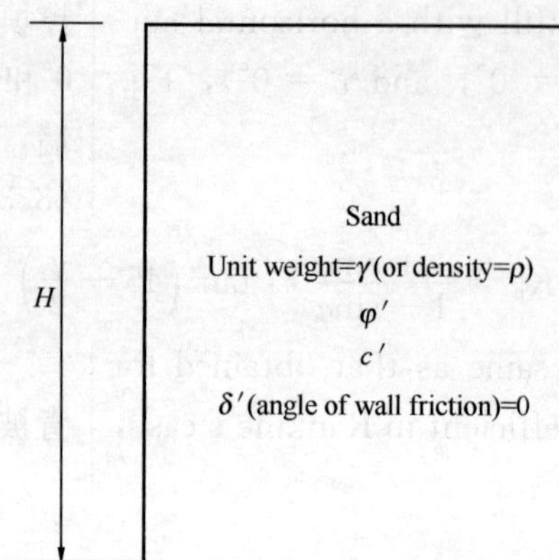

Figure 8. 19　Figure of problem 8. 3

Problem	H (m)	H_1 (m)	γ_1 (kN/m^3)	γ_2 (kN/m^3)	φ_1' (°)	φ_2' (°)	q (kN/m^2)
8.7	4.3	2.1	16.2	19.9	28	28	0
8.8	7.3	3	18	19.1	32	32	12
8.9	9	9	16.5	20.2	30	34	21

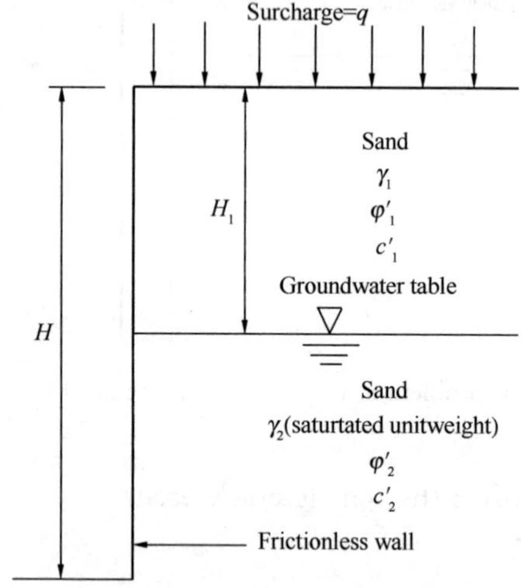

Figure 8. 20　Figure of problem 8. 4

Chapter 9 Bearing Capacity of Subsoil

第 9 章　地基承载力

9.1 Introduction

The bearing capacity of a subsoil refers to its a bility to withstand loads. A structure with certain functions consists of three parts: upper structure, foundation, and ground. The function of the subsoil is to transfer the load of the structure to the soil on which it is resting. The subsoil results in deformation under the load. As the load increases, the deformation of the subsoil gradually increases. In the initial stage, the stress in the soil is in an elastic equilibrium state and has a safe bearing capacity. When the load increases to a point or small area where the shear stress in a certain direction plane reaches the shear strength of the soil, the point or small area will undergo shear failure and be in the ultimate equilibrium state, and the stress in the soil will be redistributed. This small-scale shear failure zone is called the plastic zone. The ultimate equilibrium state of a small-scale foundation can mostly be restored to an elastic equilibrium state, and the subsoil can still tend to be stable with safe bearing capacity. When the load continues to increase and a large range of plastic zones appear in the subsoil, it will indicate insufficient bearing capacity and loss of stability. At this point, the subsoil has reached its ultimate bearing capacity.

A properly designed foundation transfers the load throughout the soil without overstressing the soil. Overstressing the soil can result in either excessive settlement or shear failure of the soil, both of which cause damage to the structure. Thus, geotechnical and structural engineers who design foundations must evaluate the bearing capacity of soils.

There are generally four methods for determining the bearing capacity of a subsoil: in-situ testing, theoretical formula method, code method, and empiri-cal method. The in-situ test method is a method of deter-

9.1 概述

地基承载力是指其承受荷载的能力。具有一定功能的结构由三部分组成：上部结构、基础和地基。基础的功能是将结构的荷载传递到其所处的地基上。地基在荷载作用下变形。随着荷载的增大，地基的变形逐渐增大。在初始阶段，土中的应力处于弹性平衡状态，具有安全承载力。当荷载增大到某一方向平面上的剪应力达到土的抗剪强度时，某点或小区域将发生剪切破坏并处于极限平衡状态，土中的应力将重新分布。这种小规模的剪切破坏区被称为塑性区。小型基础的极限平衡状态大多可以恢复到弹性平衡状态，地基仍然可以趋于稳定，具有安全承载力。当荷载继续增大，地基中出现大范围的塑性区时，表明承载力不足和稳定性丧失。此时，地基已达到其极限承载力。

设计合理的基础将荷载传递到整个土体中，而不会使土体承受过大的应力。对土体施加过大的应力会导致地基土过度沉降或剪切破坏，这两者都会对结构造成损坏。因此，设计基础的岩土和结构工程师必须评估地基承载力。

确定地基承载力的方法一般有四种：原位测试法、理论公式法、规范法和经验法。原位测试

mining the bearing capacity through direct on-site testing. The in-situ test includes (static) load test, static penetration test, standard penetration test, lateral pressure test, etc. Among them, the load test method is the most reliable in-situ test method. Theoretical formula method is a method of determining the bearing capacity based on the theoretical formula of the shear strength index of soil. The code method is a method of obtaining bearing capacity based on indoor test indicators, on-site test indicators, or field identification indicators by referring to the tables listed in the specifications. The empirical method is a method based on regional usage experience, which determines the carrying capacity through analogy judgment. It is a macroscopic auxiliary method.

法是一种通过直接现场测试来确定承载力的方法。原位测试包括（静）载荷试验、静力触探试验、标准贯入试验、旁压试验等，其中，载荷试验法是最可靠的原位试验方法。理论公式法是一种基于土抗剪强度指标的理论公式确定承载力的方法。规范法是一种根据室内试验指标、现场试验指标或现场识别指标，参考规范中列出的表格获得承载力的方法。经验法是一种基于区域使用经验的方法，通过类比判断来确定承载力。这是一种宏观的辅助方法。

In this chapter, we discuss the failure mode and the soil-bearing capacity for shallow foundations including critical load and ultimate bearing capacity.

本章，我们讨论了浅基础的破坏模式和地基承载力，包括临界荷载和极限承载力。

9.2 Ground Failure Modes of Shallow Foundation

9.2 浅基础的地基破坏模式

The real reason for the foundation failure is that the shear strength of the soil is reached. Test studies have shown that there are three types of failure modes of shallow foundation: global shear failure, local shear failure and punching shear failure shown in Figure 9.1, Figure 9.2 and Figure 9.3 respectively. To understand the concept of the ultimate soil-bearing capacity and the mode of shear failure in soil, let us consider the case of a long rectangular footing of width B located at the surface of a dense sand layer (or stiff soil). When a uniformly distributed load of p per unit area is applied to the footing, it settles. If the uniformly distributed load p is increased, the settlement of the footing gradually increases. When the value of $p=p_u$ is reached (Figure 9.1b), bearing-capacity failure occurs; the footing undergoes a very large settlement without any further increase of p. The soil on one or both sides of the foundation bulges, and the slip surface extends to the ground

地基失稳的真正原因是土体剪切强度达到极限。试验研究表明，浅基础有三种类型破坏模式：整体剪切破坏、局部剪切破坏和冲剪破坏，分别如图 9.1、图 9.2 和图 9.3 所示。为理解土体极限承载力的概念以及土中的剪切破坏方式，想象一个宽度为 B 且位于密实砂层（或坚硬土体）表面的长方形基础。当对单位面积施加均布荷载 p 时，基础发生沉降，沉降量随均布荷载 p 的增大而增大。当 $p=p_u$ 时，超出基础的承载能力，基础发生巨大的沉降，且荷载 p 将不会继续增大，如图 9.1（b）所示。基础一侧或两侧的土体隆起，滑动面延伸到地表，荷载-沉降关

surface. The load-settlement relationship is like curve shown in Figure 9.1(b). In this case, p_u is defined as the ultimate bearing capacity of subsoil. The bearing-capacity failure just described is called a general shear failure and can be explained with reference to Figure 9.1(a). When the foundation settles under the application of a load, a triangular wedge-shaped zone of soil (marked Ⅰ) is pushed down, and, in turn, it presses the zones marked Ⅱ and Ⅲ sideways and then upward. At the ultimate pressure, p_u, the soil passes into a state of plastic equilibrium and failure occurs by sliding.

系如图 9.1（b）所示，此时 p_u 被定义为地基的极限承载力。上文所描述地基承载能力破坏现象被称为整体剪切破坏，如图 9.1（a）所示。当基础在荷载作用下发生沉降时，一个三角形楔形区域的土体（标记为Ⅰ）被向下推动，进而向侧面挤压Ⅱ和Ⅲ的区域，随后向上挤压。在极限压力 p_u 作用下，土体进入塑性平衡状态并通过滑动发生破坏。

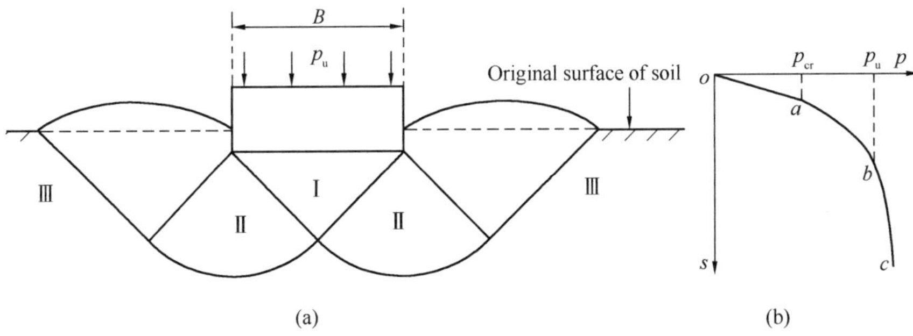

Figure 9.1 Global shear failure mode

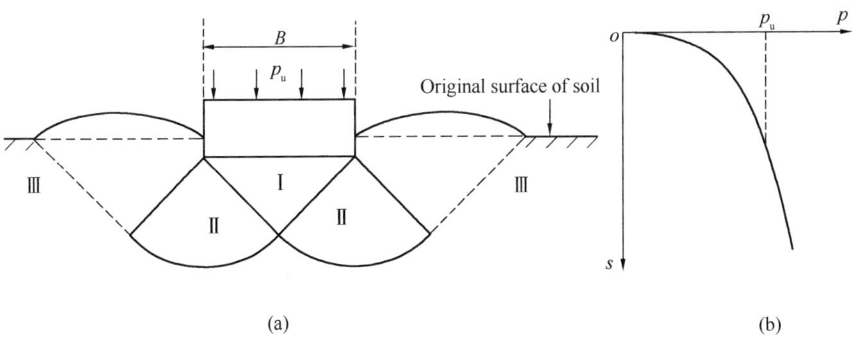

Figure 9.2 Local shear failure mode

If the footing test is conducted instead in a loose-to-medium dense sand, the load-settlement relationship is like curve in Figure 9.2(b). Beyond a certain value of $p = p_u$, the load-settlement relationship becomes a steep, inclined straight line. In this case, p_u is defined as the ultimate bearing capacity of subsoil. This type of subsoil failure is referred to as local shear failure and is

如果在松散至中等密实度的砂土中进行试验，荷载-沉降关系类似于图 9.2（b）中的曲线。当 p 值超过某个特定值 p_u 时，荷载-沉降关系变成一条陡峭的斜直线，p_u 被定义为地基的极限承载力，这种类型的地基破坏

shown in Figure 9.2(a). The triangular wedge-shaped zone (marked Ⅰ) below the footing moves downward, but unlike general shear failure, the slip surfaces end somewhere inside the soil. Some signs of soil bulging are seen, however.

被称为局部剪切破坏，如图 9.2（a）所示。基础下方的三角形楔形区域Ⅰ向下移动，但其与整体剪切破坏不同，滑动面在土体内部某处终止，土体仍存在隆起的迹象。

Punching shear failure is a type of ground failure modes in which the soil mass undergoes vertical shear failure under shallow foundation loads, resulting in significant settlement of the foundation. It is also known as piercing shear failure. Its failure characteristics are: under the action of load, the foundation undergoes significant settlement, and some of the surrounding soil also sinks. When the foundation fails, it seems to " pierce" into the soil layer of the subsoil, without obvious failure zones or sliding surfaces. The foundation has no obvious inclination, and its p-s curve has no turning point (Figure 9.3b). It is a typical failure mode characterized by deformation, as shown in Figure 9.3(a). In loose sand and soft subsoil with high compressibility, or in situations where the foundation is buried deep, shear failure is more likely to occur.

冲剪破坏是土体在浅基荷载作用下发生竖直剪切破坏，导致基础显著沉降的一种破坏模式，也被称为刺入剪切破坏。其破坏特征为：在荷载作用下，基础发生显著沉降，周围部分土体下沉，当基础失稳时，荷载似乎“刺入”了地基的土层中，没有明显的破坏带或滑动面，基础没有明显的倾斜，其 p-s 曲线无转折点如图 9.3（b）所示。这是一种典型的以变形为特征的破坏模式，如图 9.3（a）所示。在松散砂土和高压缩性的软土地基中或基础埋深较深的情况下，更容易发生冲剪破坏。

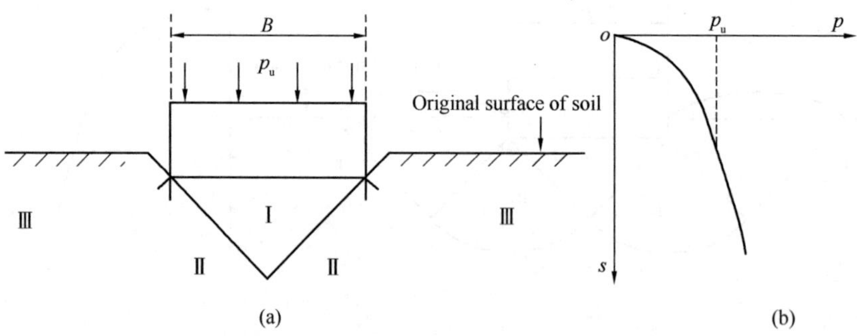

Figure 9.3 Punching shear failure mode

It must be pointed out that the failure mode of the ground is related to the conditions of the soil, such as type, density, moisture content, compressibility, shear strength, and other factors. The conditions of foundation, such as form, burial depth, size, etc., among which the compressibility of soil is the main factor affecting the failure mode. If the compressibility of the

必须指出，地基的破坏模式不仅与土体条件有关，如土体类型、密度、含水量、压缩性、剪切强度等因素，同时也与基础条件相关，如基础形状、埋深、大小等，其中土体压缩性是影响破坏模式的主要因素。如果土体的

soil is low and the soil is relatively dense, it is generally prone to general shear failure. On the contrary, if the soil is relatively loose and has high compressibility, it is prone to punching shear failure.

压缩性差且相对密实，通常较易发生整体剪切破坏。相反，若土体相对松散且具有高压缩性，则容易发生冲剪破坏。

9.3 Critical Load of Subsoil

9.3 地基临界荷载

The results of on-site loading test indicate that when the subsoil undergoes general shear failure, there are three distinct stages on its p-s curve (Figure 9.1b): compression stage, shear stage, and heave stage shown in Figure 9.4.

现场载荷试验结果表明，当地基发生整体剪切破坏时，其 p-s 曲线（图 9.1b）上有三个明显不同的阶段：压缩阶段、剪切阶段和隆起阶段，见图 9.4。

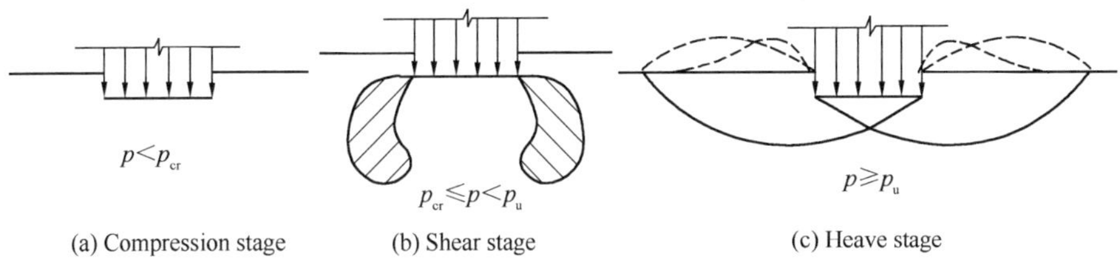

Figure 9.4 Three stage of stress state in subsoil

(1) The compression stage, also known as the linear deformation stage, corresponds to the *oa* (Figure 9.1b) segment of the p-s curve. At this stage, the external load is relatively small, and the subsoil mainly undergoes compression deformation. The relationship between pressure and deformation is basically linear. At this time, the stress in the foundation is still in an elastic equilibrium state, and the shear stress at any point in the foundation is less than the shear strength at that point. The stress at this stage can generally be analyzed approximately using elastic theory.

（1）压缩阶段，也称为线性变形阶段，对应 p-s 曲线的 *oa* 段（图 9.1b）。此阶段外部荷载相对较小，地基土体主要发生压缩变形，压力与变形之间的关系近似呈线性。此时，地基中应力仍处于弹性平衡状态，任意点的剪切应力都小于该点的剪切强度。此阶段的应力一般可用弹性理论进行近似分析。

(2) The shear stage, also known as the plastic deformation stage, corresponds to the *ab* segment of the p-s curve. At this stage, starting from the bottom edges on both sides of the foundation, the shear stress in the local soil area is equal to the shear strength of the soil at that location, and the soil undergoes plastic deformation. Macroscopically, the p-s curve shows nonlinear changes. As the load increases, the plastic deformation zone of the soil under the foundation ex-

（2）剪切阶段，也称为塑性变形阶段，对应 p-s 曲线的 *ab* 段。此阶段在基础两侧底部边缘开始，局部土体区域的剪切应力等于该处土体的剪切强度，此区域发生塑性变形，宏观上 p-s 曲线呈非线性变化。随荷载增大，基础下方土体塑性变形区域扩大，荷载变形曲线斜率增大，

pands, and the slope of the load deformation curve increases. At this stage, although some areas of the subsoil undergo plastic deformation, the plastic deformation zone is not connected to the foundation, and the foundation still has a certain degree of stability. The safety of the foundation decreases as the plastic deformation zone expands.

在这一阶段，尽管地基土体局部区域发生塑性变形，但塑性变形区未与基础相连，基础仍具有一定的稳定性。随着塑性变形区不断扩大，基础的安全性逐渐降低。

(3) The heave stage, also known as the plastic flow stage, corresponds to the *bc* segment of the *p-s* curve. At this stage, the plastic deformation zones on both sides of the foundation below the foundation are connected and linked together, and the soil on both sides of the foundation rises. In this stage, even small load increments can cause large settlement of the foundation. This deformation is mainly caused by the plastic flow of the subsoil, which is an unstable deformation over time. The result is that the foundation tilts to one side, and the overall stability of the foundation is lost.

（3）隆起阶段，也称为塑性流动阶段，对应 *p-s* 曲线的 *bc* 段。在这一阶段，基础下方两侧的塑性变形区域连接并贯通，基础两侧的土体上升，此时，即使施加轻微荷载也将导致基础大幅度沉降。这种变形主要由地基土体的塑性流动引起，是随时间推移的不稳定变形。最终表现为基础向一侧倾斜，整体稳定性丧失。

There are two boundary loads corresponding to the three stages of stress state in the subsoil: the first is the boundary load that transitions from the compression stage to the shear stage, called proportional limit loading or critical edge loading, generally referred to as p_{cr}, which is the load corresponding to point *a* on the *p-s* curve; The latter is the boundary load that transitions from the shear stage to the heave stage, called the ultimate loading, denoted as p_u, which is the load corresponding to point *b* on the *p-s* curve. Generally, the characteristic value of the bearing capacity of shallow foundations is determined by using p_{cr} or p_u/K (where *K* is the safety factor).

地基土体中三个应力状态阶段对应着两个临界荷载：第一个是从压缩阶段过渡到剪切阶段的边界荷载，称为比例界限荷载或临界边缘荷载，通常简称为 p_{cr}，这是 *p-s* 曲线上点 *a* 对应的荷载；第二个是从剪切阶段过渡到隆起阶段的临界荷载，称为极限荷载，表示为 p_u，这是 *p-s* 曲线上点 *b* 对应的荷载。通常，浅基础承载力的特征值通过 p_{cr} 或 p_u/K（*K* 是安全系数）来确定。

9.3.1 Boundary Equation of Plastic Deformation Zone

According to the theory of elasticity, the major and minor principal stresses generated at any point *M* in the soil under uniformly distributed strip loads (Figure 9.5) can be expressed as:

9.3.1 塑性变形区的边界方程

根据弹性理论，在均匀分布的条形荷载作用下土体中任意点 *M* 处产生的大、小主应力（图 9.5），可以表示为式（9.1）。

$$\begin{matrix}\sigma_1 \\ \sigma_3\end{matrix} = \frac{p-\gamma_m d}{\pi}(\beta_0 \pm \sin\beta_0) \tag{9.1}$$

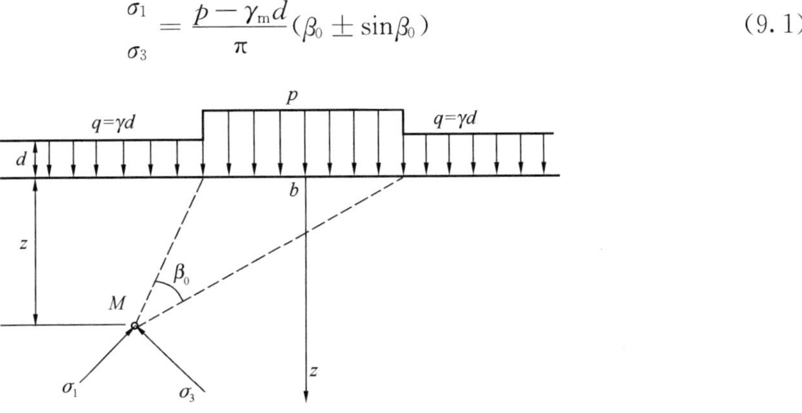

Figure 9.5 Principal stress in soil under uniformly distributed strip load

where p is base pressure; γ_m is the average unit weight of soil above the foundation bottom; d is the burial depth of the foundation; β_0 is the angle between any point M and the two endpoints of a uniformly distributed strip load.

式中，p 为基底压力；γ_m 为基础底部以上土体的平均重度；d 为基础的埋深；β_0 为任意点 M 与均布条形荷载两端点的夹角。

The direction of action of σ_1 at point M is consistent with the bisector of angle β_0. The stress acting on point M, in addition to the additional stress on the foundation caused by the average additional pressure p_0 on the foundation, also includes the self-weight stress of the soil, $q+\gamma z$. γ is the unit weight of the bearing layer soil.

σ_1 在点 M 的作用方向与角 β_0 的平分线一致。作用于点 M 上的应力，除基础平均附加压力 p_0 对基础产生的附加应力外，还包括土的自重应力 $q+\gamma z$，其中 γ 是持力层土的重度。

Assuming that the at-rest lateral pressure coefficient K_0 of the original self-weight stress field of the subsoil is equal to 1.0 and has the property of hydrostatic pressure, that is, $\sigma_x=\sigma_z=q+\gamma z$, the self-weight stress field does not change the magnitude of the additional stress field at point M and the direction of the principal stress. Therefore, the major and minor principal stresses at any point M in the foundation are

假设地基土原自重应力场的静止侧压力系数 K_0 等于 1.0，且具有静水压力的性质，即 $\sigma_x=\sigma_z=q+\gamma z$，则自重应力场在点 M 处和主应力方向的附加应力场的大小不发生变化。因此，地基中任何一点 M 处的大、小主应力可表达为式（9.2）。

$$\begin{matrix}\sigma_1 \\ \sigma_3\end{matrix} = \frac{p-\gamma_m d}{\pi}(\beta_0 \pm \sin\beta_0) + q + \gamma z \tag{9.2}$$

When point M reaches the limit equilibrium state, the major and minor principal stresses at that point should satisfy the limit equilibrium conditions. Therefore, according to the limit equilibrium condition, it can

当点 M 达到极限平衡状态时，该点处的大、小主应力应满足极限平衡条件。因此，根据极限平衡条件，可从式（9.2）得

be obtained from Eq. (9.2) that

式（9.3）。

$$z=\frac{(p-\gamma_m d)}{\pi\gamma}\left(\frac{\sin\beta_0}{\sin\varphi}-\beta_0\right)-\frac{1}{\gamma}(c\cot\varphi+q) \tag{9.3}$$

The above equation is the boundary equation of the plastic zone in the soil under a certain pressure p. From this equation, it can be inferred that when the physical and mechanical properties of the soil (γ, c, φ), base pressure, and burial depth are known, the coordinate z of any point on the boundary of the plastic zone is a function of β_0. According to this formula, the boundary line of the plastic zone can be drawn as shown in Figure 9.6.

上述方程是在一定压力 p 下土中塑性区的边界方程。可以推断，当已知土体的物理力学性质（γ、c、φ）、基底压力和埋深时，塑性区边界上任意点的坐标 z 是 β_0 的函数。根据该式，可绘制出如图 9.6 所示的塑性区边界线。

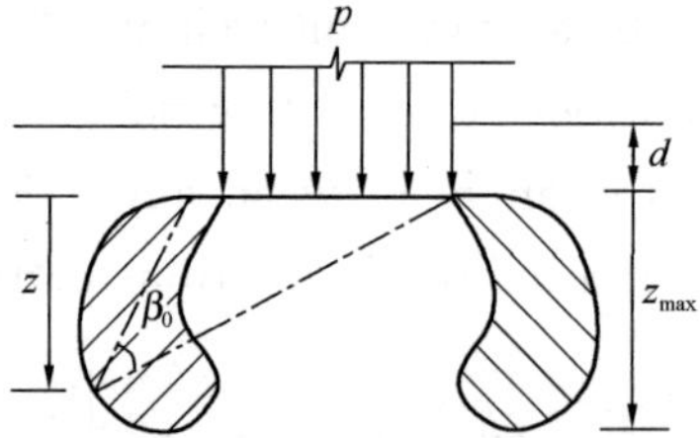

Figure 9.6 Plastic zone at the edge of the bottom of foundation

9.3.2 Critical Edge Load and Critical Loading

According to Eq. (9.3), the depth z of the plastic zone boundary is a function of β_0. By taking the first derivative of function z with respect to β_0 and making it equal to 0, it can be obtained that

9.3.2 临塑荷载和临界荷载

根据式（9.3），塑性区边界的深度 z 是 β_0 的函数。通过取 z 相对于 β_0 的一阶导数并令其等于 0，可得有关 β_0 的表达式。

$$\frac{dz}{d\beta_0}=\frac{p-\gamma_m d}{\pi\gamma}\left(\frac{\cos\beta_0}{\sin\varphi}-1\right)=0$$

$$\beta_0=\frac{\pi}{2}-\varphi$$

By substituting β_0 into Eq. (9.3), the expression for the maximum depth z_{max} of the plastic zone boundary can be derived as

将 β_0 代入式（9.3），可推导出塑性区边界的最大深度 z_{max} 的表达式，见式（9.4）。

$$z_{max}=\frac{p-\gamma_m d}{\pi\gamma}\left(\cot\varphi-\frac{\pi}{2}+\varphi\right)-\frac{1}{\gamma}(c\cot\varphi+q) \tag{9.4}$$

As can be seen from the above equation, when the load p_0 ($p-\gamma_m d$) increases, the plastic zone develops and expands, and the maximum depth of the plastic zone also increases. According to the definition, the critical edge load is the load when the soil is about to enter the plastic zone, that is, the load when $z_{max}=0$.

由上式可得，当荷载 p_0（$p-\gamma_m d$）增大时，塑性区域扩大，最大深度增加，根据定义可知，临塑荷载为土体即将进入塑性区的荷载，即 $z_{max}=0$ 时的荷载，令式（9.4）右侧为零，可

If the right-hand side of Eq (9.4) is zero, the formula for the critical edge load p_{cr} can be obtained as

得临塑荷载 p_{cr} 的表达式为式(9.5a)或式(9.5b)。

$$p_{cr} = \frac{\pi(c\cot\varphi + \gamma_m d)}{\cot\varphi + \varphi - \pi/2} + \gamma_m d \tag{9.5a}$$

or

$$p_{cr} = cN_c + qN_q \tag{9.5b}$$

where N_c and N_q are called bearing capacity coefficients, both of which are functions of φ.

式中，N_c和 N_q被称为承载力系数，它们都是 φ 的函数。

$$N_c = \pi\cot\varphi/(\cot\varphi + \varphi - \pi/2)$$
$$N_q = (\cot\varphi + \varphi + \pi/2)/\tau(\cot\varphi + \varphi - \pi/2)$$
$$q = \gamma_m d$$

From Eqs. (9.5a) and (9.5b), it can be seen that the critical edge load p_{cr} includes two parts, one is provided by the cohesive force of the soil, and the other is the effect of the foundation burial depth.

由式(9.5a)和式(9.5b)可见，临界边缘荷载 p_{cr} 包括两部分，一部分由土的黏聚力提供的，另一部分受基础埋深的影响。

When designing foundation engineering, if the critical edge load p_{cr} that does not allow the subsoil to produce plastic zones is used as the characteristic value of the soil bearing capacity, it often cannot fully utilize the bearing capacity of the soil, and the value tends to be conservative. The critical load of the plastic zone in the controllable foundation within a certain depth range can be used as the characteristic value of the subsoil bearing capacity, so that the subsoil has sufficient safety and stability, and can fully exert its bearing capacity, thereby achieving the purpose of optimizing design, reducing the amount of subsoil engineering, and saving investment, which is in line with the principle of economic rationality.

在基础工程的设计中，如果将不允许地基产生塑性变形区对应的荷载（临界荷载）p_{cr} 作为土的承载力特征值，往往不能充分利用土的承载能力，且数值趋于保守。在一定深度范围内，可将可控地基塑性区的临界荷载作为地基的承载力特征值，使地基有足够的安全性和稳定性，并充分发挥其承载力，从而达到优化设计、减少地基工程量、节约投资的目的，符合经济合理的原则。

According to engineering practice, it is permissible to control the maximum depth of plastic zone development $z_{max}=b/4$ under central load and $z_{max}=b/3$ under eccentric load for the safety of general buildings. b is the foundation width of buildings. Then define $p_{1/4}$ and $p_{1/3}$ as the two critical loads corresponding to the plastic zones that allow the subsoil to generate $z_{max}=b/4$ and z_{max} $b/3$, respectively. According to this definition, substituting $z_{max}=b/4$ and $z_{max}=b/3$ into Eq. (9.4) yields

根据工程实践，为确保一般建筑物的安全，在中心荷载作用下允许控制塑性区扩大的最大深度为 $z_{max}=b/4$，在偏心荷载作用下的最大深度为 $z_{max}=b/3$，其中 b 为建筑物基础宽度。则定义 $p_{1/4}$和 $p_{1/3}$分别为允许地基产生 $z_{max}=b/4$ 和 $z_{max}b/3$ 的塑性区所对应的两种临界荷载。按此定义，将 $z_{max}=b/4$、$z_{max}=b/3$ 分别代入式（9.4）得式（9.6）和式（9.7）。

$$p_{1/4}=cN_c+qN_q+\gamma bN_{1/4} \tag{9.6}$$

$$p_{1/3}=cN_c+qN_q+\gamma bN_{1/3} \tag{9.7}$$

where $N_{1/4}$ and $N_{1/3}$ are called bearing capacity coefficients, both of which are functions of φ.

式中，$N_{1/4}$ 和 $N_{1/3}$ 称为承载力系数，均为 φ 的函数。

$$N_{1/4}=\pi/[4(\cot\varphi+\varphi-\pi/2)]$$

$$N_{1/3}=\pi/[3(\cot\varphi+\varphi-\pi/2)]$$

According to Eqs. (9.6) and (9.7), it can be seen that compared with the critical edge load p_{cr}, the critical load consists of three parts, which increases the effect of foundation width and subsoil unit weight.

由式（9.6）和式（9.7）可见，与临塑荷载 p_{cr} 相比，临界荷载由三部分组成，增加了基础宽度和地基土重度的影响。

It must be pointed out that the formulas for critical edge load and critical load are derived under strip load conditions (plane strain problems). For rectangular or circular foundations (spatial problems), using this formula for calculation may yield results that are biased towards safety. In the derivation of critical loads $p_{1/4}$ and $p_{1/3}$, a plastic zone has already appeared in the subsoil, but it is still calculated using elastic mechanics formulas. The error caused by this increases as the plastic zone expands. In addition, during the derivation process, it is assumed that the at-rest soil pressure coefficient $K_0=1$, and the distribution of uniformly distributed strip loads differs significantly from actual engineering, which also brings errors to the calculation results.

必须指出，临塑荷载和临界荷载公式是在条形荷载下（平面应变问题）推导的，对于矩形或圆形基础（空间问题）按此式计算可能产生偏于安全性的结果；在临界荷载 $p_{1/4}$ 和 $p_{1/3}$ 推导中，地基土已出现塑性区，但仍继续使用弹性力学公式进行计算，由此产生的误差将随塑性区的扩大而增大；此外，在推导过程中假定静止土压力系数 $K_0=1$，均布条形荷载的分布与实际工程有较大的差异，也会产生计算误差。

Example 9.1

A certain strip foundation is placed on a homogeneous soil, with a width of 3.0m and a burial depth of 1.0m. The natural unit weight of the subsoil is 18.0 kN/m^3, the natural moisture content is 38%, the relative density of soil particles is 2.73, the shear strength index $c=15$kPa, and $\varphi=12°$. Estimate p_{cr}, $p_{1/4}$ and $p_{1/3}$. If the groundwater level rises to the bottom of the foundation, assuming that the shear strength index of

例题 9.1

某条形基础置于均质地基上，宽度 $b=3.0$m，埋深 $d=1.0$m，地基土天然重度 $\gamma=18.0$kN/m^3，天然含水量 $w=38\%$，土粒相对密度 $d_s=2.73$，抗剪强度指标 $c=15$kPa，$\varphi=12°$，试估算 p_{cr}、$p_{1/4}$ 和 $p_{1/3}$，若地下水位上升到基础地面，假

the soil remains constant, what changes will be made to p_{cr}, $p_{1/4}$ and $p_{1/3}$?

Solution

According to $\varphi=12°$, it can be obtained $N_c=4.42$, $N_q=1.94$, $N_{1/4}=0.23$, $N_{1/3}=0.31$. $q=\gamma_m d=18.0\times1.0=18.0\text{kPa}$. Substituting them into Eqs. (9.5b), (9.6) and (9.7) respectively, it can be obtained that

设土的抗剪强度指标不变，则 p_{cr}、$p_{1/4}$和 $p_{1/3}$有何变化？

解答

根据 $\varphi=12°$，可查得 $N_c=4.42$，$N_q=1.94$，$N_{1/4}=0.23$，$N_{1/3}=0.31$；$q=\gamma_m d=18.0\times1.0=18.0\text{kPa}$，分别代入式（9.5b）、式（9.6）和式（9.7），可得 p_{cr}、$p_{1/4}$和 $p_{1/3}$的表达式。

$$p_{cr}=cN_c+qN_q=15\times4.42+18.0\times1.94=101\text{kPa}$$

$$p_{1/4}=cN_c+qN_q+\gamma bN_{1/4}=15\times4.42+18.0\times1.94+18.0\times3.0\times0.23=114\text{kPa}$$

$$p_{1/3}=cN_c+qN_q+\gamma bN_{1/3}=15\times4.42+18.0\times1.94+18.0\times3.0\times0.31=118\text{kPa}$$

If the groundwater level rises to the bottom of the foundation, at this time, γ should be replaced by buoyant unit weight γ'.

当地下水位上升到基础地面时，此时 γ 需取浮重度为 γ'，可得 p_{cr}、$p_{1/4}$和 $p_{1/3}$的表达式。

$$\gamma'=\frac{(d_s-1)\gamma}{d_s(1+w)}=\frac{(2.73-1)\times18.0}{2.73\times(1+0.38)}=8.27\text{kN/m}^3$$

so

$$p_{cr}=15\times4.42+18.0\times1.94=101\text{kPa}$$

$$p_{1/4}=15\times4.42+18.0\times1.94+8.27\times3.0\times0.23=107\text{kPa}$$

$$p_{1/3}=15\times4.42+18.0\times1.94+8.27\times3.0\times0.31=109\text{kPa}$$

As can be seen from comparison, $p_{cr}<p_{1/4}<p_{1/3}$. When the groundwater level rises, there is no change in the critical edge load value, while the critical loading value decreases. Therefore, the rise of groundwater level will lead to a decrease in the bearing capacity of the subsoil.

由比较可见，$p_{cr}<p_{1/4}<p_{1/3}$，当地下水位上升后，临界边缘荷载值不变，而临界荷载值减小，因此地下水位的上升将导致地基承载力下降。

9.4 Ultimate Bearing Capacity of Subsoil

The ultimate subsoil bearing capacity refers to the ultimate load that subsoil can withstand when it is about to become unstable due to shear failure, also known as the ultimate load of subsoil. It is equivalent to the boundary load when the stress state in the subsoil transitions from the shear stage to the heave stage. The load value corresponding to point b on the p-s curve.

There are many theoretical formulas for the ulti-

9.4 地基极限承载力

地基极限承载力是指地基即将发生剪切破坏失稳时的极限荷载，也称地基的极限荷载，相当于地基土应力状态从剪切阶段过渡到隆起阶段的界限荷载，即 p-s 曲线上 b 点对应的荷载值。

地基极限承载力的理论公式

mate bearing capacity of subsoil, most of which are derived based on the general shear failure mode, and are modified based on experience when used for local shear or punching shear failure situations.

较多，大多是根据整体剪切破坏模式推导出来的，对于局部剪切和冲剪破坏情况则需根据经验进行修正。

In this section, we discuss the Prandtl and Reissner's ultimate bearing capacity equations, Terzaghi's ultimate bearing capacity equation and general bearing capacity equation.

本节讨论普朗特和赖斯纳极限承载力公式、太沙基极限承载力公式和一般承载力公式。

9.4.1 Prandtl and Reissner's Ultimate Bearing Capacity Equations

9.4.1 普朗特和赖斯纳极限承载力公式

In 1921, Prandtl published the results of his study on the penetration of hard bodies (such as metal punches) into a softer material. Prandtl assumes that an infinitely long, smooth bottomed rigid foundation is placed on a homogeneous, isotropic, and zero gravity subsoil surface. When the load acting on the foundation is large enough, the subsoil undergoes overall shear failure, and the subsoil below the foundation is in a plastic equilibrium state, with the plastic zone divided into five zones as shown in Figure 9.7.

普朗特于 1921 年发表了有关坚硬物体（如金属块）压入较软的半无限体材料的研究结果，假设将无限长、底面光滑的刚性基础置于均匀、各向同性、零重力地基表面上。当作用于基础的荷载足够大时，地基发生整体剪切破坏，基础下土体处于塑性平衡状态，塑性区分为 5 个区域，如图 9.7 所示。

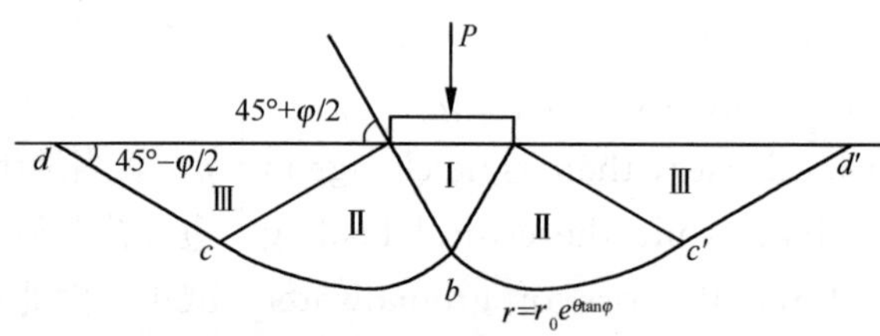

Figure 9.7 Prandtl's general shear failure mode of subsoil

Ⅰ zone is the central wedge located below the foundation bottom, also known as the Rankine active zone. The direction of action of the major principal stress σ_1 in this zone is vertical, and the direction of action of the minor principal stress σ_3 is horizontal. According to the limit equilibrium theory, the direction of action of the minor principal stress is at an angle of $(45°+\varphi/2)$ with the failure surface, that is, the angle between the two sides of the central zone and the horizontal plane. Adjacent to the central area are two shear zones, also known

Ⅰ区是基础底面下的中央楔形区，即朗肯主动区，该区内大主应力 σ_1 的作用方向是垂直的，小主应力 σ_3 为水平向，根据极限平衡理论，小主应力作用方向与破坏面呈 $(45°+\varphi/2)$ 夹角，即中央区两侧与水平面夹角；与中央区相邻的为两个剪切区，即 Prandtl 区，由一组对数螺旋线和一组辐射直线组成，该区形状

as Prandtl zones, consisting of a set of logarithmic helices and a set of radiating straight lines. This zone is shaped like a sector with logarithmic helices $r_0 e^{(\theta \tan\varphi)}$ as the arc-shaped boundary, and its central angle is a right angle. Adjacent to the other side of the Prandtl zone is the Rankine passive zone, where the direction of the major principal stress is horizontal and the direction of the small principal stress σ_3 is vertical. The angle between the fracture surface and the horizontal plane is $(45°-\varphi/2)$. Prandtl derived the expression for the ultimate load acting on the foundation bottom under the above conditions, that is, the equation for the ultimate bearing capacity is

像扇形，以对数螺旋线 $r_0 e^{(\theta \tan\varphi)}$ 为弧形边界，其中心角为直角。在 Prandtl 区另一侧是朗肯被动区，大主应力作用方向为水平，小主应力 σ_3 作用方向为垂直，破坏面与水平面的夹角为 $(45°-\varphi/2)$。普朗特根据以上条件推导出作用于基础底面的极限荷载公式，即极限承载力公式(9.8)，

$$p_u = cN_c \tag{9.8}$$

where N_c is bearing capacity coefficients.

式中，N_c 为承载力系数。

$$N_c = \cot\varphi\left[e^{(\pi\tan\varphi)} \tan^2(45° + \varphi/2) - 1\right]$$

In 1924, Ressiner took the Prandtl zone as the object of force analysis, its force condition is shown in Figure 9.8.

1924 年，赖斯纳取普朗特区为受力分析对象，其受力情况如图 9.8 所示。

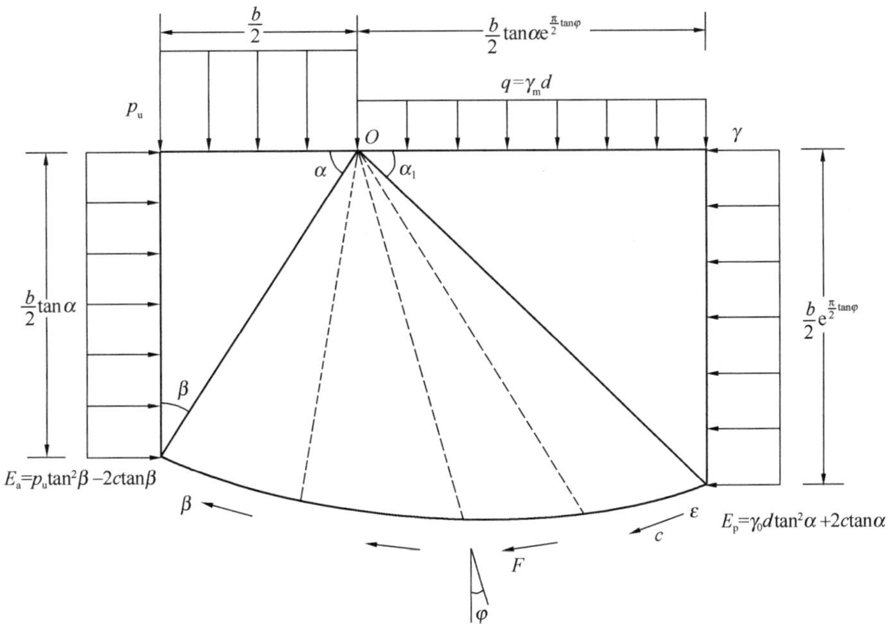

Figure 9.8 Prandtl's bearing-capacity analysis

In the mechanism shown in the Figure 9.8, the combined moment of each force at point O is 0. From this, it can be obtained that

图 9.8 中各力对 O 点的合力矩为 0，由此可得各力的表达式。

$$M_1 = \frac{1}{8} p_u b^2$$

$$M_2 = \frac{1}{8} p_u b^2 - \frac{c}{4} b^2 \tan\alpha$$

$$M_3 = \frac{1}{8} \gamma_0 d b^2 e^{\pi\tan\varphi} \tan^2\alpha + \frac{1}{4} b^2 \cdot c \cdot e^{\pi\tan\varphi} \tan\alpha$$

$$M_4 = \frac{1}{8} b^2 \gamma_0 d e^{\frac{\pi}{2}\tan\varphi} \tan^2\alpha$$

$$M_5 = \frac{1}{8} b^2 c \frac{e^{\pi\tan\varphi} - 1}{\tan\varphi\cos^2\alpha}$$

$$M_1 + M_2 - M_3 - M_4 - M_5 = 0$$

Therefore,

因此可知荷载 p_u 的表达式为式（9.9），式中 N_c 为承载力系数。

$$p_u = cN_c + qN_q \tag{9.9}$$

where

$$N_q = e^{(\pi\tan\varphi)} \tan^2(45° + \varphi/2)$$

9.4.2 Terzaghi's Ultimate Bearing Capacity Equations

In 1943, Terzaghi extended the plastic failure theory of Prandtl to evaluate the bearing capacity of soils for shallow strip footings considering the following conditions.

(1) Unit weight of the subsoil $\gamma \neq 0$;

(2) The bottom of foundation is rough;

(3) Ignore the shear strength of the fill above the foundation, it is only regarded as an overload acting on the horizontal plane of the foundation;

(4) Under the action of ultimate load, the subsoil undergoes overall shear failure;

(5) Assuming the shape of the sliding surface in the subsoil is as shown in Figure 9.9.

9.4.2 太沙基极限承载力公式

1943 年，太沙基将普朗特塑性破坏理论推广到浅基础承载力公式中，并考虑以下条件：

（1）地基土重度 $\gamma \neq 0$；

（2）基底粗糙；

（3）忽略基础上方填土的抗剪强度，仅视为作用在基础水平面上的超载；

（4）在极限荷载作用下，地基发生整体剪切破坏；

（5）假设地基滑动面形状如图 9.9 所示。

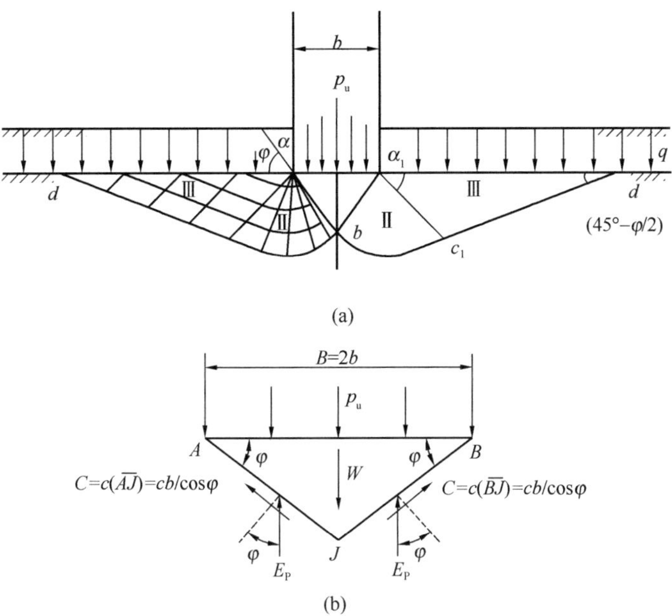

Figure 9.9 Terzaghi's bearing-capacity analysis

If the load per unit area p_u, is applied to the footing and general shear failure occurs, the passive force E_p is acting on each of the faces of the soil wedge ABJ. This concept is easy to conceive of if we imagine that AJ and BJ are two walls that are pushing the left and right soil wedges Ⅱ in Figure9-9(a), respectively, to cause passive failure. E_p should be inclined at an angle δ (which is the angle of wall friction) to the perpendicular drawn to the wedge faces (that is, AJ and BJ). In this case, δ should be equal to the angle of friction of soil, φ. Because AJ and BJ are inclined at an angle φ to the horizontal, the direction of E_p should be vertical.

如果将单位面积荷载 p_u作用于基底，并发生整体剪切破坏，则被动土压力 E_p将作用于土楔 ABJ 的任何面。如果我们想象 AJ 和 BJ 是两面墙，它们分别推动图 9.9(a) 中左、右侧土楔 Ⅱ，从而导致被动破坏。E_p应该与垂直于楔面（即 AJ 和 BJ）的垂线成 δ 角（δ 是墙摩擦角）。在这种情况下，δ 应等于土的内摩擦角 φ。由于 AJ 和 BJ 与水平面成 φ 角倾斜，因此 E_p的方向应该是竖直的。

Now let us consider the free-body diagram of the wedge ABJ as shown in Figure 9.9(b). Considering the unit length of the footing, we have, for equilibrium,

现在让我们考虑一下楔形 ABJ 的自由体，如图 9.9(b) 所示。考虑基础的单位长度，根据平衡，可得式 (9.10)。

$$p_u \times 2b \times 1 = -W + 2C\sin\varphi + 2E_p \tag{9.10}$$

where $b=B/2$; W is weight of soil wedge ABJ, $W=\gamma b^2\tan\varphi$; C is cohesive force acting along each face, AJ and BJ, that is equal to the unit cohesion times the length of each face, $cb/\cos\varphi$.

式中，$b=B/2$；$W=\gamma b^2\tan\varphi$ 为土楔 ABJ 的重量；C 为施加在每个作用面 AJ 和 BJ 上的黏聚力，等于单位黏聚力乘以作用面

的长度，$cb/\cos\varphi$。因此可得有关p_u的表达式（9.11）和式（9.12）

Thus,

$$2bp_u = 2E_p + 2bc\tan\varphi - \gamma b^2\tan\varphi \tag{9.11}$$

or

$$p_u = \frac{E_p}{b} + c\tan\varphi - \frac{\gamma b}{2}\tan\varphi \tag{9.12}$$

The passive pressure in Eq.（9.11）is the sum of the contribution of the unit weight of soil γ, cohesion c, and surcharge q and can be expressed as

式（9.11）中的被动土压力是土的重度γ、黏聚力c和超载q贡献的总和，可以表示为式（9.13）。

$$E_p = c(b\tan\varphi)K_c + q(b\tan\varphi)K_q + \frac{1}{2}\gamma(b\tan\varphi)^2K_\gamma \tag{9.13}$$

where K_c, K_q, and K_γ are earth pressure coefficients that are functions of the soil friction angle, φ.

其中K_c、K_q和K_γ为土压力系数，它们是有关土内摩擦角φ的函数。

Combining Eqs.（9.12）and（9.13），we obtain

将式（9.12）和式（9.13）结合，可得到p_u有关N_c，N_q和N_γ的表达式，

$$p_u = cN_c + qN_q + \frac{1}{2}\gamma bN_\gamma$$

where

$$N_c = \tan\varphi(K_c + 1) \tag{9.14}$$

$$N_q = K_q\tan\varphi \tag{9.15}$$

$$N_\gamma = \frac{1}{2}\tan\varphi(K_\gamma\tan\varphi - 1) \tag{9.16}$$

The terms N_c, N_q, and N_γ are, respectively, the contributions of cohesion, surcharge, and unit weight of soil to the ultimate load-bearing capacity. It is extremely tedious to evaluate K_c, K_q, and K_γ. For this reason, Terzaghi used an approximate method to determine the ultimate bearing capacity, p_u. The principles of this approximation are the following.

式中，N_c，N_q和N_γ分别为黏聚力、超载和土的重度对极限承载力的影响。估算K_c、K_q和K_γ极其繁琐。因此，太沙基选取一种近似方法确定极限承载力p_u，此方法原则如下：

（1）If $c=0$ and surcharge $q=0$（that is, $d_f=0$），then

（1）当$c=0$，且超载$q=0$（即$d_f=0$）时，p_u为土自重施加的承载力，式（9.17）。

$$p_u = p_\gamma = \frac{1}{2}\gamma bN_\gamma \tag{9.17}$$

（2）If $\gamma=0$（that is, weightless soil）and $q=0$,

（2）当$\gamma=0$（即无重量的

then

土）且 $q=0$，则 p_u 为黏聚力施加的承载力，式（9.18）。

$$p_u = p_c = cN_c \tag{9.18}$$

(3) If $\gamma=0$ (weightless soil) and $c=0$, then

（3）当 $\gamma=0$（即无重量的土）且 $c=0$，则 p_u 为超载施加的承载力，式（9.19）。

$$p_u = p_q = qN_q \tag{9.19}$$

By the method of superimposition, when the effects of the unit weight of soil, cohesion, and surcharge are considered, we have

通过叠加法，当考虑土的重度、黏聚力和超载的影响时，可得式（9.20）。

$$p_u = p_c + p_q + p_\gamma = cN_c + qN_q + \frac{1}{2}\gamma bN_\gamma \tag{9.20}$$

Equation (9.20) is referred to as Terzaghi's bearing-capacity equation. The terms N_c, N_q, and N_γ are called the bearing-capacity factors. The values of these factors are given in Table 9.1.

式（9.20）被称为太沙基极限承载力公式。N_c，N_q和 N_γ被称为承载力系数，其值可由表9.1确定。

Terzaghi's Bearing-Capacity Factors **Table 9.1**

φ (°)	N_c	N_q	N_γ	φ (°)	N_c	N_q	N_γ
0	5.70	1.00	0.00	26	27.09	14.21	9.84
1	6.00	1.10	0.01	27	29.24	16.90	11.60
2	6.30	1.22	0.04	28	31.61	17.81	13.70
3	6.62	1.35	0.06	29	34.24	19.98	16.18
4	6.97	1.49	0.10	30	37.16	22.46	19.13
5	7.34	1.64	0.14	31	40.41	25.28	22.65
6	7.73	1.81	0.20	32	44.04	28.52	26.87
7	8.15	2.00	0.27	33	48.09	32.23	31.94
8	8.60	2.21	0.35	34	52.64	36.50	38.04
9	9.09	2.44	0.44	35	57.75	41.44	45.41
10	9.61	2.69	0.56	36	63.53	47.16	54.36
11	10.16	2.98	0.69	37	70.01	53.80	65.27
12	10.76	3.29	0.85	38	77.50	61.55	78.61
13	11.41	3.63	1.04	39	85.97	70.61	95.03
14	12.11	4.02	1.26	40	95.66	81.27	116.31
15	12.86	4.45	1.52	41	106.81	93.85	140.51
16	13.68	4.92	1.82	42	119.67	108.75	171.99
17	14.60	5.45	2.18	43	134.58	126.50	211.56
18	15.12	6.04	2.59	44	161.95	147.74	261.60
19	16.56	6.70	3.07	45	172.28	173.28	325.34
20	17.69	7.44	3.64	46	196.22	204.19	407.11
21	18.92	8.26	4.31	47	224.55	241.80	512.84
22	20.27	9.19	5.09	48	258.28	287.85	650.67
23	21.75	10.23	6.00	49	298.71	344.63	831.99
24	23.36	11.40	7.08	50	347.50	416.14	1072.80
25	25.13	12.72	8.34				

Problems

9.1 A continuous footing is shown in Figure 9.10. Using Terzaghi's bearing-capacity factors, determine the gross allowable load per unit area (q_{all}) that the footing can carry. Assume general shear failure. Given: $\gamma=17.5\text{kN/m}^3$, $c=21\ \text{kN/m}^2$, $\varphi=32°$, $d=1\text{m}$, $B=1.5\text{m}$, and factor of safety is 3.

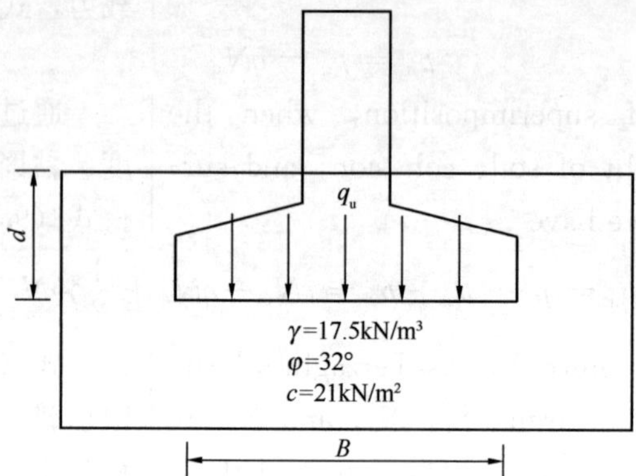

Figure 9.10 Figure of problem 9.1

9.2 Repeat Problem 9.1 with the following: $\gamma=19.5\text{kN/m}^3$, $c_u=37\text{kN/m}^2$, $\varphi=0$, $d=0.75\text{m}$, $B=2.5\text{m}$, and factor of safety is 6.

Chapter 10 Slope Stability

第 10 章 边坡稳定性

10. 1 Introduction

10. 1 概述

An exposed ground surface that stands at an angle with the horizontal is called an unrestrained slope. The slope can be natural or man-made. When both the top and bottom surfaces of a soil slope are horizontal and extend to infinity, and are composed of homogeneous soil, it is called a simple soil slope as shown in Figure 10. 1.

无约束边坡是指与水平面成一定角度的裸露地面，边坡可以是自然形成的，也可以是人工建造的。当土坡的顶面和底面均为水平向且向两侧无限延伸，同时由均质土体构成时，称为简单土坡，如图 10. 1 所示。

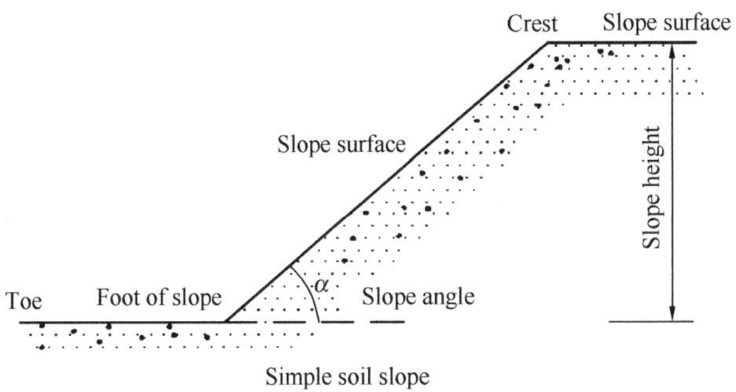

Figure 10. 1 Simple soil slope

Due to the inclined surface of the soil slope, the soil will exhibit a top-down sliding trend under its own weight and external loads. The phenomenon of partial rock or soil mass on a soil slope undergoing shear failure and moving downwards along a clear interface under the influence of natural or human factors is called a slide or slope failure as shown in Figure 10. 2.

由于土坡表面倾斜，在自重及外荷载的作用下，土体具有自上而下的滑动趋势。如图 10. 2 所示，土坡内部分岩土体在自然或人为因素的影响下，发生剪切破坏并沿明确界面向下运动的现象称为滑坡或边坡破坏。

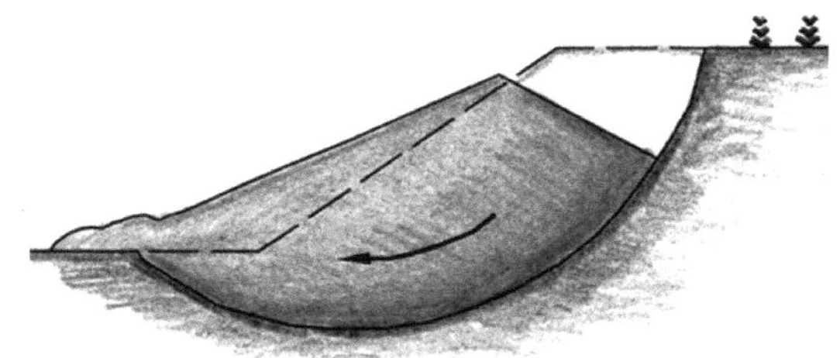

Figure 10. 2 Slope sliding failure

The essence of soil slope sliding instability is that the shear stress on a certain sliding surface inside the

土坡滑动失稳的实质是土体内部某一滑动面的剪应力达到其

soil reaches its shear strength, causing the stability balance to be disrupted. The factors that induce soil slope sliding instability are complex and varied, but generally can be classified into two categories. One is the increase in shear stress inside the soil caused by external loads or changes in the soil slope environment, such as excavation of road cuts or foundation pits, increase in the load of upper fill soil during dam construction, increase in soil saturation caused by rainfall, increase in water permeability inside the soil, excessive load at the top of the slope, or dynamic loads caused by earthquakes, pile driving, etc. Another type is due to various external factors that lead to a decrease in the shear strength of soil, causing instability and failure of soil slopes, such as an increase in pore water pressure, dry cracking and freeze-thaw caused by climate change, softening of clay interlayers due to rainwater intrusion, and a decrease in soil strength caused by creep of cohesive soil.

抗剪强度，导致稳定平衡被破坏。引起土坡滑动失稳的因素复杂多变，一般可分为两类：一类是由于外部荷载或土坡环境变化引起的土体内部剪应力增大，如路堑或基坑开挖、筑坝过程中上部填方土荷载增大、降雨引起的土体饱和增大、土体内部透水性增大、边坡顶部荷载过大或地震、打桩等引起的动荷载；另一类是由于各种外部因素而导致的土体抗剪强度降低，引起土坡失稳和破坏，如气候变化引起孔隙水压力增大、干裂和冻融，雨水侵入引起黏土夹层软化，黏性土蠕变引起土体强度降低。

Slope sliding failure often occurs in civil engineering constructions such as highways, railways, airports, deep excavation of high-rise buildings, as well as open-pit mines and earth dams. The sliding instability of slopes is an engineering hazard that requires engineers to evaluate the stability of slopes and determine the possibility of sliding failure, and then take feasible engineering measures to prevent and control it.

边坡滑动破坏常发生在公路、铁路、机场、高层建筑深基坑以及露天矿、土坝等土木工程建设中。边坡的滑动失稳是一种工程灾害，需要工程师对边坡的稳定性进行评估，确定发生滑动破坏的可能性，并采取可行的工程措施进行预防和控制。

In this chapter, we will discuss in detail the definition of factor of safety, stability of non-cohesive soil slopes and stability of cohesive soil slopes.

在本章中，我们将详细讨论安全系数的定义、无黏性土边坡的稳定性和黏性土边坡的稳定性。

10.2 Factor of Safety

10.2 安全系数

The task of the engineer charged with analyzing slope stability is to determine the factor of safety. Generally, the factor of safety is defined as

分析边坡稳定性的主要任务是确定安全系数。安全系数通常定义为式（10.1），

$$K = \frac{\tau_f}{\tau_d} \tag{10.1}$$

where K is the factor of safety with respect to strength, τ_f is the average shear strength of the soil, τ_d is the average shear stress developed along the potential failure surface.

式中，K 为相对于强度的安全系数，τ_f 为土体的平均抗剪强度，τ_d 为沿潜在破坏面发展的平均剪应力。

The shear strength of a soil consists of two components, cohesion and friction, and may be written as

土的抗剪强度由黏聚力和摩擦力两个分量组成，可以写为式(10.2)，

$$\tau_f = c + \sigma\tan\varphi \tag{10.2}$$

where c is cohesion, φ is angle of friction, σ is normal stress on the potential failure surface.

In a similar manner, we can write

式中，c 为黏聚力；φ 为摩擦角；σ 为潜在破坏面的法向应力。同样地，我们可以写为式（10.3），

$$\tau_d = c_d + \sigma\tan\varphi_d \tag{10.3}$$

where c_d and φ_d are, respectively, the cohesion and the angle of friction that develop along the potential failure surface. Substituting Eqs. (10.2) and (10.3) into Eq. (10.1), we get

式中，c_d和 φ_d分别是沿潜在破坏面发展的黏聚力和摩擦角。将式(10.2) 和式 (10.3) 代入式(10.1) 中，可得到式（10.4）。

$$K = \frac{c + \sigma\tan\varphi}{c_d + \sigma\tan\varphi_d} \tag{10.4}$$

Now we can introduce some other aspects of the factor of safety, that is, the factor of safety with respect to cohesion, K_c, and the factor of safety with respect to friction, K_φ. They are defined as

接下来我们介绍安全系数的其他方面，即关于黏聚力的安全系数 K_c 和摩擦的安全系数 K_φ，定义为式（10.5）和式（10.6）。

$$K_c = \frac{c}{c_d} \tag{10.5}$$

and

$$K_\varphi = \frac{\tan\varphi}{\tan\varphi_d} \tag{10.6}$$

When we compare Eqs. (10.4) through (10.6), we can see that when K_c becomes equal to K_φ, it gives the factor of safety with respect to strength. Or, if

比较式（10.4）～式(10.6) 可知，当 $K_c = K_\varphi$ 时，给出了相对于强度的安全系数，或者如果满足等式，

$$\frac{c}{c_d} = \frac{\tan\varphi}{\tan\varphi_d}$$

then we can write

则可以写成式（10.7）。

$$K = K_c = K_\varphi \tag{10.7}$$

When K is equal to 1, the slope is in a state of impending failure. Generally, a value of no more than 1.5 for the factor of safety with respect to strength is acceptable for the design of a stable slope.

当安全系数 $K=1$ 时，边坡处于即将失稳状态。一般来说，对于稳定边坡的设计，安全系数与强度的比值不小于 1.5 属于可接受范围。

10.3 Stability of Non-cohesive Soil Slopes

Assuming a homogeneous and non-cohesive soil slope with a slope of β as shown in Figure 10.3(a), the slope body and its foundation are of the same type of soil, and are completely dry or completely submerged, that is, there is no seepage effect. Due to the lack of cohesion between particles in non-cohesive soil, as long as the soil unit located on the slope can remain stable, the entire soil slope is stable.

10.3 无黏性土边坡的稳定性

假设边坡由均质无黏性土构成，如图 10.3（a）所示，坡度为β，坡体和地基由相同类型土构成，且完全干燥或完全饱和，即不存在渗流效应。由于无黏性土颗粒间缺乏黏聚力，只要边坡上的土单元能够保持稳定，即视为整个边坡稳定。

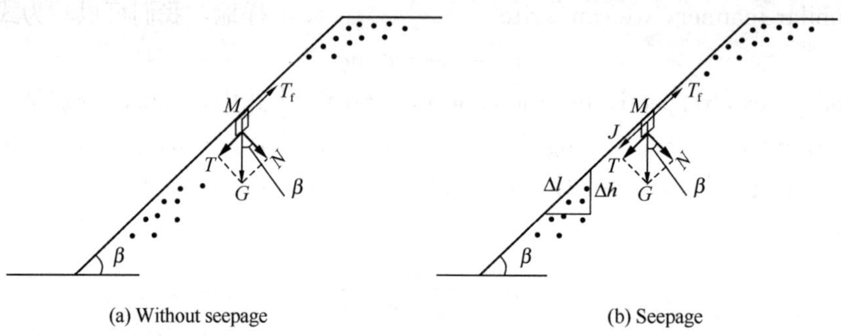

Figure 10.3 Analysis of non-cohesive soil slopes

On any slope, take a soil element M with one side vertical and the bottom parallel to the slope. Without considering the influence of stress on stability on both sides of the element, the self-weight of the element is G, and the internal friction angle of the soil is φ. Therefore, the shear force that causes the soil element to slide down is the component force of G in the direction of the slope, $T=G\sin\beta$. The force that prevents the soil from sliding down is the shear resistance force T_f between the unit cell and the underlying soil, which is equal to the frictional force caused by the component force N of the unit cell's own weight in the normal direction of the slope surface, i. e. $T_f=N\tan\varphi=G\cos\beta\tan\varphi$. The ratio of anti-skid force to sliding force is called the stability safety factor, represented by K, that is

在任意斜坡上，取一土单元M，其侧面竖直，底部与坡面平行，忽略土单元两侧应力对稳定性的影响，单元自重记为G，土的内摩擦角记为φ。土单元沿坡面下滑的剪切力由G的水平分量提供，表达式为$T=G\sin\beta$。阻止土体向下滑动的力是土单元与下层土之间的剪切阻力T_f，该阻力等于土单元自重垂直分量N在坡面法向上的摩擦力$T_f=N\tan\varphi=G\cos\beta\tan\varphi$。抗滑力与滑动力的比值称为稳定安全系数，可表示为式（10.8）。

$$K=\frac{T_f}{T}=\frac{G\cos\beta\tan\varphi}{G\sin\beta}=\frac{\tan\varphi}{\tan\beta} \tag{10.8}$$

As can be seen from the above, for homogeneous

由上可知，对于均质无黏性

and non-cohesive soil slopes, theoretically the stability of the soil slope is independent of the slope height. As long as the slope angle is smaller than the internal friction angle of the soil ($\beta<\varphi$) and $K>1.0$, the soil is stable. When the slope angle is equal to the internal friction angle of the soil ($\beta=\varphi$), the stability safety factor $K=1.0$, and the anti-sliding force is equal to the sliding force, the soil slope is in the ultimate equilibrium state, and the corresponding slope angle is equal to the internal friction angle of the non-cohesive soil, which is called the natural angle of repose. Usually, in order to ensure that the soil slope has sufficient safety reserve, $K\geqslant1.3\sim1.5$ can be taken.

土边坡，土体的稳定性主要取决于其坡角与土的内摩擦角关系，而与坡高无关，只要坡角小于土的内摩擦角（$\beta<\varphi$）且 $K>1.0$，边坡即处于稳定状态。当坡角达到土的内摩擦角（$\beta=\varphi$），此时安全系数 $K=1.0$，抗滑力与滑动力达到平衡，边坡处于极限平衡状态，此时的坡角称为无黏性土的自然休止角。为保证土质边坡具有足够的安全性，通常取 $K\geqslant1.3\sim1.5$。

When a non-cohesive soil slope is subjected to a certain amount of seepage force, the unit soil on the slope where the seepage overflows is not only affected by its own weight, but also by the seepage force $J=\gamma_w i$ (where i is the hydraulic gradient and $i=\sin\beta$), as shown in Figure 10.3 (b). If the seepage flow is along the slope, the direction of seepage flow and seepage force at the overflow point is parallel to the slope surface. At this time, the shear force of the soil unit sliding down is $T+J=G\sin\beta+\gamma_w i$, and for the unit soil, the self-weight G of the soil is equal to the effective weight γ'. Therefore, the stability safety factor of the soil slope becomes

无黏性土边坡存在渗流作用时，渗流溢出边坡上的单位土体不仅受到自身重力的影响，还受到渗流力 $J=\gamma_w i$（其中 i 为水力梯度，$i=\sin\beta$）的影响，如图 10.3（b）所示。如果渗流沿边坡流动，则溢流点处的渗流方向和渗流力与边坡表面平行。此时，土体单元下滑的剪力为 $T+J=G\sin\beta+\gamma_w i$，对于单位土，土体的自重 G 等于有效重量 γ'，土边坡的稳定性安全系数可表示为式（10.9），

$$K=\frac{T_f}{T+J}=\frac{\gamma'\cos\beta\tan\varphi}{(\gamma'+\gamma_w)\sin\beta}=\frac{\gamma'}{\gamma_{sat}}\frac{\tan\varphi}{\tan\beta} \tag{10.9}$$

It can be seen that compared with Eq. (10.8), there is a difference of γ'/γ_{sat} times, which is about 1/2. Therefore, when there is seepage along the slope, the stability safety factor of non-cohesive soil slopes is reduced by about half.

相较于式（10.8）相差 γ'/γ_{sat}，约为1/2。因此，渗流会导致无黏性土边坡的稳定性安全系数约降低至原来的一半。

10.4 Stability of Cohesive Soil Slopes

10.4 黏性土边坡的稳定性

The instability mode of cohesive soil slopes is related to engineering geological conditions. For homogeneous cohesive soil slopes, the sliding surfaces that fail due to shear are mostly curved surfaces of homogeneous

黏性土边坡的失稳模式受工程地质条件的影响。在均质黏性土边坡中，剪切破坏导致的滑动面通常呈现为曲线形态。破坏

soil slopes. Before failure, tension cracks usually appear at the top of the slope, and then overall sliding occurs along a certain curved surface. In addition, the sliding body also has a certain range along the longitudinal direction and is also a curved surface. In order to simplify the stability analysis, it is often assumed that the sliding surface is a cylindrical surface and treated as a plane strain problem. According to the different slope angles, soil strength indicators, and the position of the hard layer in the soil, there are generally three forms of cylindrical sliding surfaces: the circular sliding surface passing through point B at the foot of the slope (Figure 10.4a) is called the foot circle; The circular sliding surface passing through point E on the slope surface (Figure 10.4b) is called the slope surface circle; The circular sliding surface passing through point A outside the foot of the slope (Figure 10.4c) is called the midpoint circle.

前，边坡顶部常出现张拉裂缝，而后边坡沿特定曲线面发生整体滑移。滑移体在纵向上也延伸一定范围，形成曲面。为简化分析，常将滑动面假定为圆柱形，并按平面应变问题处理。根据坡角、土体强度指标以及土体内硬层的位置，圆柱形滑动面主要有三种形式：穿过边坡底部 B 点的圆弧面（图 10.4a）称为底圆；穿过坡面 E 点的圆弧面（图 10.4b）称为坡圆；穿过边坡底部外侧 A 点的圆弧面（图 10.4c）称为中点圆。

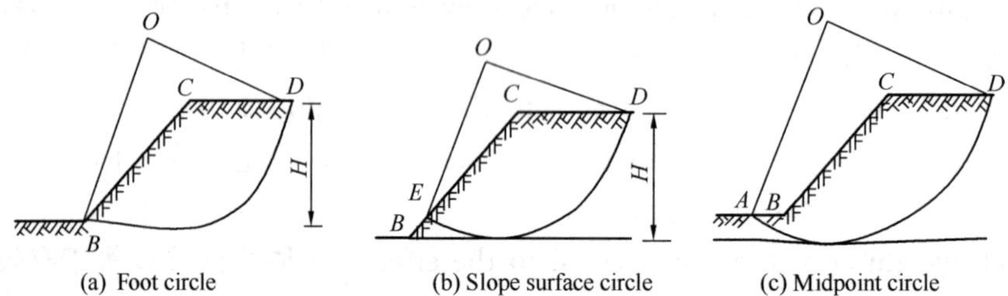

Figure 10.4　Three types of circular sliding surfaces for homogeneous cohesive soil slopes

10.4.1　Analysis of Finite Slopes with Circular Failure Surfaces-General

If the simple soil slope is homogeneous, the failure surface can be assumed to be a circular arc in section. The soil above the sliding surface is regarded as a rigid body, and it is used as a detachment body to analyze the various forces acting on the detachment body under limit equilibrium conditions. The safety factor can be calculated according to the definition in Section 10.1.

Only moment equilibrium is considered in this section. A trial failure surface (centre O, radius R and length L_a) is shown in Figure 10.5. Potential instability

10.4.1　有限边坡的常规圆弧滑动面分析

若简单土坡为均质土体，则可假定其破坏面在截面上为圆弧。将滑动面上土体视为刚体，并将其作为分离体，分析在极限平衡条件下作用在分离体上的各种力。安全系数根据第 10.1 节中的定义计算。

本节只考虑力矩平衡，试验破坏面（圆心 O，半径 R，长度 L_a）如图 10.5 所示，潜在的不

is due to the total weight of the soil mass (G per unit length) above the failure surface. The driving (clockwise) moment about O is therefore $M_A = Gd$. The soil resistance is described by an anticlockwise moment $M_R = \tau_f L_a R$ about O. If the slip surface is a circular arc, then $L_a = R\theta$ from Figure 10. 5. The criterion of stability at ULS (ultimate limit states) is then described by $M_A \leqslant M_R$.

稳定性是由于破坏表面上方的土体总重量（单位长度为 G）引起的。因此，以 O 为圆心的驱动力矩（顺时针方向）为 $M_A = Gd$。土的阻力用关于 O 的逆时针力矩 $M_R = \tau_f L_a R$ 来表示。如果滑动面为圆弧，则图 10. 5 中的 $L_a = R\theta$。稳定性的极限状态（ULS）准则可表示为 $M_A \leqslant M_R$。

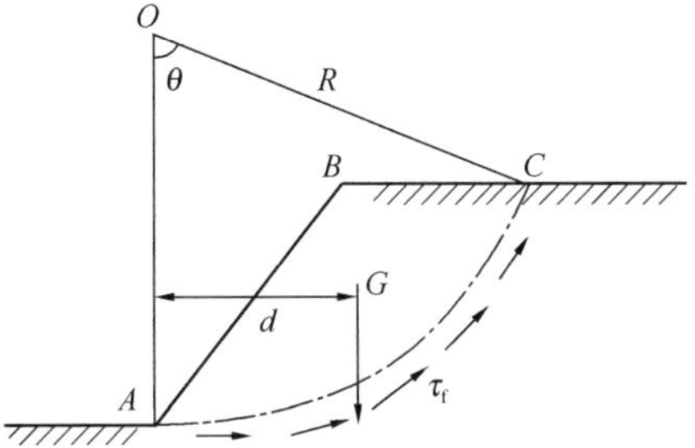

Figure 10. 5 Limit equilibrium analysis in homogeneous soil

The moments of any additional forces (e. g. surcharge) must be taken into account in determining M_A. In the event of a tension crack developing, the arc length L_a is shortened and a hydrostatic force will act normal to the crack if it fills with water. It is necessary to analyze the slope for a number of trial failure surfaces in order that the most critical failure surface can be determined. In the analysis of an existing slope, M_A will be less than M_R (as the slope is standing) and the safety of the slope is usually expressed as a factor of safety K where

在确定土坡稳定性时，需考虑所有额外力附加力对力矩（M_A）的影响。在发生拉伸裂缝的情况下，弧长 L_a 会缩短，如果裂缝充满水，静水力将垂直于裂缝。为确定最关键的破坏面，需要对多个试验破坏面进行边坡分析。在分析现有边坡时，M_A 将小于 M_R（因为边坡是站立的），并且边坡的安全通常表示为安全系数 K，可表示为式（10. 10）。

$$K = \frac{M_R}{M_A} = \frac{\tau_f L_a R}{Gd} = \frac{\tau_f R\theta R}{Gd} = \frac{\tau_f R^2 \theta}{Gd} \tag{10.10}$$

From Eq. (10. 10) it can be seen that a stable slope will have $K > 1.0$, while an unstable slope will have $K < 1.0$. If $K = 1.0$, the slope is at the point of failure and the ULS stability criterion is regained.

由式（10. 10）可知，稳定边坡 $K > 1.0$，不稳定边坡 $K < 1.0$，当 $K = 1.0$ 时，边坡处于破坏点，并且恢复 ULS 稳定性准则。

The shear strength of soil is composed of two parts: cohesive force c and frictional force $\sigma\tan\varphi$. The

土的抗剪强度由黏聚力 c 和摩擦力 $\sigma\tan\varphi$ 组成。土体中法向

normal stress σ in soil is not constant along the sliding surface, so the shear strength of soil also varies with the position of the sliding surface. But for saturated clay, under undrained shear conditions, $\varphi_u=0$, so $\tau_f=c_u$. Therefore, the above equation can be written as

应力 σ 沿滑动面不是恒定的，因此土体的抗剪强度也随滑动面的位置而变化。而对于饱和黏土，在不排水剪切条件下 $\varphi_u=0$，所以 $\tau_f=c_u$，因此，上述等式可以写为式（10.11）。

$$K=\frac{M_R}{M_A}=\frac{\tau_f L_a R}{Gd}=\frac{c_u R^2\theta}{Gd} \tag{10.11}$$

Due to the assumption that the sliding surface is arbitrary and not the most dangerous sliding surface when calculating the above safety factor, the result obtained is not the minimum safety factor. Usually, a series of sliding surfaces need to be assumed during calculations, and multiple trials need to be conducted, resulting in a considerable amount of computational workload. Therefore, in 1927, Fellenius proposed an empirical method for determining the center of the most dangerous sliding surface through extensive computational analysis.

在计算上述安全系数时，由于假设滑动面是任意的，并不是最危险的滑动面，所以得到的结果并不是最小安全系数。在计算时通常需要假设一系列滑动面，并且需要进行多次试验，计算量相当大。因此，在 1927 年，费伦纽斯提出了一种经验方法，通过广泛的计算分析来确定最危险的滑动面的中心。

Fellenius believed that when the internal friction angle of soil is equal to 0, its most dangerous sliding surface often passes through the foot of the slope. The center position can be determined by the intersection of the *CO* and *BO* lines in Figure 10.6(a), and the values of β_1 and β_2 in the figure can be found from Table 10.1 based on the slope angle. When $\varphi>0$, the center position of the most dangerous sliding surface may be on the extension line of *EO* in Figure 10.6(b). Take the center points O_1, O_2... from point *O* to the outside, and draw the sliding arc passing through point *C*. Calculate the corresponding anti sliding safety factors K_1, K_2..., and then draw a curve to find the minimum value, which is the required center point O_m of the most dangerous sliding surface and the stability safety factor K_{min} of the soil slope. When the soil slope is heterogeneous or the slope shape and load situation are complex, it is necessary to draw a vertical line from O_m as the *OE* line, and take several points on the vertical line as the center of the circle for calculation and compari-

费伦纽斯理论认为，在土体内摩擦角 $\varphi=0$ 的情况下，边坡最危险的滑动面通常穿过坡脚，圆心位置可由图 10.6（a）中 *CO* 和 *BO* 线的交点确定，图中 β_1 和 β_2 的值可根据坡角从表 10.1 中得到。当 $\varphi>0$ 时，最危险的滑动面中心位置可能在图 10.6（b）中 *EO* 延长线上，取中心点 O_1、O_2... 从 *O* 点向外，画出通过 *C* 点的滑动弧。计算相应的抗滑安全系数 K_1，K_2……然后画出曲线求最小值，即最危险滑动面所需圆心点 O_m 和土坡稳定安全系数 K_{min}。当土质边坡不均匀或边坡形状和荷载情况复杂时，需要从 O_m 处画一条垂直线作为 *OE* 线，并以垂直线上的几个点作为圆心进行计算和比较，以找到最危险滑动面的

son, in order to find the center of the most dangerous sliding surface and the stability safety factor of the soil slope.

中心和土质边坡的稳定安全系数。

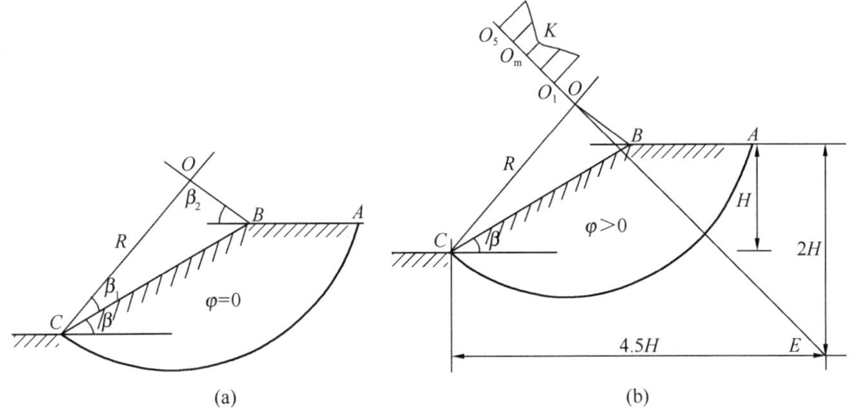

Figure 10.6 Determination of the center position of the most dangerous sliding surface

β_1 and β_2 **Table 10.1**

Slope ratio	Slope angle	β_1	β_2
1：0.58	60°	29°	40°
1：1	45°	28°	37°
1：1.5	33.79°	26°	35°
1：2	26.57°	25°	35°
1：3	18.43°	25°	35°
1：4	14.04°	25°	37°
1：5	11.32°	25°	37°

10.4.2 Swedish Slice Method

The Swedish slice method is the oldest and simplest slice method, also known as the Swedish circle method. Except for assuming that the sliding surface is a cylindrical surface and the sliding soil is a rigid body without deformation, the forces on both sides of the soil slice are ignored, so the number of unknowns is (n+1). Then, using the balance of the normal force on the bottom surface of the soil slice and the balance of the moment of the entire sliding soil slice, the expressions for the magnitude of the normal force N_i on the bottom surface of each soil slice and the stability safety factor K of the soil slope are obtained.

Assuming the sliding surface of a homogeneous soil

10.4.2 瑞典条分法

瑞典条分法，也称瑞典圆法，是最古老、最简单的条分法。该法假设滑动面为圆柱形，并且滑动土体为无变形刚体，忽略土条两侧的力，故未知数为(n+1)。然后利用土条底面法向力的平衡和整个滑动土条弯矩的平衡，得到各土条底面法向力的大小 N_i 和土坡稳定安全系数 K 的表达式。

假设均质土坡的滑动面为

slope is AC, the center of the circle is O, the radius is R, and the sliding soil ABC is divided into several soil slices as shown in Figure 10.7. If any of the soil slices (the i-th slice) is taken to analyze its stress, the forces acting on the soil slice are as follows:

AC，圆心为 O，半径为 R，将滑动土 ABC 分成若干土条，如图 10.7 所示。取任一土条（第 i 条）进行应力分析，则作用在土条上的力为：

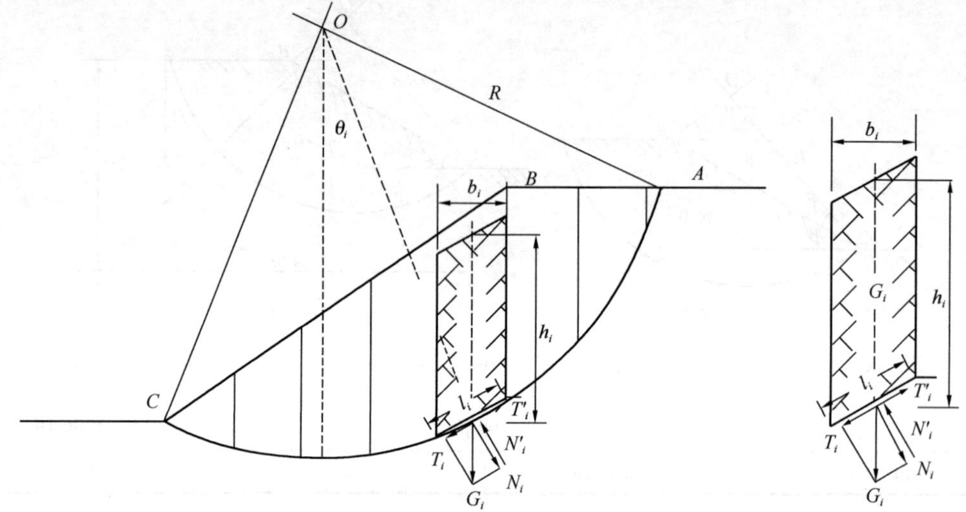

Figure 10.7 Stability analysis by Swedish slice method

(1) The self-weight G_i of the soil slice is vertically downward, with a value of

（1）土条自重 G_i 竖直向下，其值为式（10.12）。

$$G_i = \gamma b_i h_i \tag{10.12}$$

where γ is the unit weight of the soil, and b_i and h_i are the width and average height of the soil slice, respectively. Guiding G_i onto the sliding surface can be decomposed into a normal force N_i passing through the center of the sliding arc; And the shear force T_i tangent to the sliding arc. If θ_i is used to represent the angle be tween the normal of the midpoint of the bottom surface of the soil slice and the vertical line, then there is

式中，γ 为土的重度，b_i 和 h_i 分别为土条的宽度和平均高度。将 G_i 引导到滑动面上可分解为两个分量：法向力 N_i 穿过滑动弧中心；剪切力 T_i 与滑动弧相切。如果用 θ_i 表示土条底表面中点法线与垂直线的夹角，则有 N_i，T_i。

$$N_i = G_i \cos\theta_i$$

$$T_i = G_i \sin\theta_i$$

(2) Normal force N_i and reactive force N'_i acting on the bottom surface of the soil slice; Equal in size, opposite in direction.

（2）作用于土条底面的法向力 N_i 和反力 N'_i 大小相等，方向相反。

(3) Acting on the bottom surface of the soil T_{fi}, the maximum possible shear force is equal to the product of the shear strength of the soil on the bottom surf-

（3）作用于土底面 T_{fi} 时，最大可能剪力等于土条底面土抗剪强度与滑动弧长度的乘积，方

ace of the soil slice and the length of the sliding arc, with the direction opposite to the sliding direction. When the soil slope is in a stable state and assuming that the safety factor on the sliding surface at the bottom of each soil slice is equal to the safety factor on the entire sliding surface, its shear resistance is

向与滑动方向相反。当土坡处于稳定状态时，假设每条土条底部滑动面上的安全系数等于整个滑动面上的安全系数，其剪切阻力为式（10.13）。

$$T_{fi} = \frac{cl_i + N_i \tan\varphi}{K} = \frac{cl_i + G_i \cos\alpha_i \tan\varphi}{K} \tag{10.13}$$

If the moment balance of each soil slice within the entire sliding soil is taken with respect to the center O, then

若取整个滑动土体内各土条相对于中心 O 的力矩平衡，则有式（10.14），

$$\sum T_i R = \sum T_{fi} R \tag{10.14}$$

Therefore, the safety factor is

安全系数表示为式（10.15）。

$$K = \frac{\sum (cl_i + G_i \cos\theta_i \tan\varphi)}{\sum G_i \sin\theta_i} \tag{10.15}$$

If the width of each soil slice is equal, the above equation can be simplified as

如果各土条宽度相等，则上式可简化为式（10.16），

$$K = \frac{cL + \gamma b \tan\varphi \sum h_i \cos\theta_i}{\gamma b \sum h_i \sin\theta_i} \tag{10.16}$$

where L is the length of sliding arc.

In addition, attention should be paid to the position of the soil slices when calculating. When the center of the bottom surface of the soil slice is on the right side of the perpendicular line to the center O of the sliding arc, the direction of the shear force T_i is the same as the sliding direction, exerting a shear effect and taking the positive sign; When the center of the bottom surface of the soil slice is on the left side of the perpendicular line to the center of the circle, the direction of T_i is opposite to the sliding direction, playing a shear resistance role and taking a negative sign. Regardless of the location, the direction of T_{fi} is always opposite to the sliding direction.

Assuming different sliding arcs, different K values can be obtained, where the minimum K value is the stability safety factor of the soil slope. If the effective stress method is used for analysis, the actual shear resistance exerted by the bottom of the soil slice is

式中，L 为滑动弧长度。

另外，计算时要注意土条的位置：当土条底面中心位于与滑动弧圆心 O 垂直线的右侧时，剪力 T_i 方向与滑动方向相同，产生剪切作用，取正值；土条底面中心在与圆心垂直线左侧时，T_i 方向与滑动方向相反，起抗剪切作用，取负号。无论在什么位置，T_{fi} 的方向总是与滑动方向相反。

假设不同的滑动弧可以得到不同的 K 值，其中最小的 K 值为土坡的稳定安全系数。如采用有效应力法进行分析，则土条底部实际施加的剪切阻力为式（10.17），

$$T_{fi}=\frac{[c'+(\sigma_i-u_i)\tan\varphi']l_i}{K}=\frac{(G_i\cos\theta_i-u_il_i)\tan\varphi'+c'l_i}{K} \tag{10.17}$$

Therefore, the safety factor is

因此，安全系数为式（10.18），

$$K=\frac{\sum[(G_i\cos\theta_i-u_il_i)\tan\varphi'+c'l_i]}{\sum G_i\sin\theta_i}=\frac{\sum[(\gamma b_ih_i\cos\theta_i-u_il_i)\tan\varphi'+c'l_i]}{\sum\gamma b_ih_i\sin\theta_i} \tag{10.18}$$

where c', φ' are the effective stress intensity index of soil; u_i is pore water pressure at the center point of the bottom surface of the i-th soil slice.

式中，c'，φ'是土的有效应力强度指标；u_i为第i土条底面中心点孔隙水压力。

Example 10.1

A homogeneous cohesive soil slope with a height of 20m, a slope ratio of 1∶2, a fill unit weight of 18.0kN/m^3, a cohesion c of 10.0kPa, and an internal friction angle of 20°. Determine the safety factor of the soil slope using the Swedish slice method.

例题 10.1

一均质黏性土边坡高度为20m，坡度比为1∶2，填土重度为18.0kN/m^3，黏聚力c为10.0kPa，内摩擦角为20°。采用瑞典条分法确定土坡的安全系数。

Solution

(1) Select the center of the sliding arc and create the corresponding sliding arc. Draw a soil slope profile at a certain proportion (Figure 10.8). Due to the homogeneous soil slope, it can be found from Table 10.1 that $\beta_1=25°$, $\beta_2=35°$, and the intersection point O of the BO and CO lines is obtained. Find point E again and draw an extension line for EO. Take a point O_1 on this extension line as the center of the sliding arc for the first trial calculation. Draw the corresponding sliding arc at the foot of the slope and measure its radius R to be 40m.

(2) Divide the sliding soil into several soil slices and number them. For the convenience of calculation, the width b of the soil slice is taken as equal width of $0.2R=8$m. The number of soil slices starts from the perpendicular line at the center of the sliding arc and is 0. The soil slices in the reverse sliding direction are 0, 1, 2, 3…, and the soil slices in the forward sliding direction are −1, −2, −3…

(3) Measure the center height h_i of each soil slice and calculate $\sin\theta_i$, $\cos\theta_i$ and $\sum h_i\sin\theta_i$, $\sum h_i\cos\theta_i$; equivalent values in a list (Table 10.2). It should also

解答

（1）选择滑动弧的圆心，创建相应的滑动弧。绘制一定比例的土坡剖面图（图 10.8）。由于土质边坡是均质的，由表 10.1 可知$\beta_1=25°$，$\beta_2=35°$，得到BO线与CO线交点O。再找到E点，画出EO的延伸线。在这条延长线上取点O_1作为滑动弧的圆心，进行第一次试算。在坡脚处画出相应的滑动弧，测量其半径R为40m。

（2）将滑动土划分为若干土条并编号。为方便计算，取宽度$b=0.2R=8$m 的等宽土条。土条数从滑动弧心的垂直线开始，为0。反向滑动方向的土条为0、1、2、3……，正向滑动方向的土条为−1、−2、−3……

（3）测量各土条的中心高度h_i，计算$\sin\theta_i$，$\cos\theta_i$，$\sum h_i\sin\theta_i$和$\sum h_i\cos\theta_i$（表 10.2）。需要注

be noted that when taking equal width, the width of the soil slices at both ends of the soil may not be exactly equal to b. In this case, the actual height of the soil slices needs to be converted to the height corresponding to b, and $\sin\theta$ should also be calculated based on the actual width.

意在取等宽时，土体两端土条的宽度可能不完全等于 b，此时需要将土条的实际高度换算为 b 对应的高度，同时也要根据实际宽度计算 $\sin\theta$。

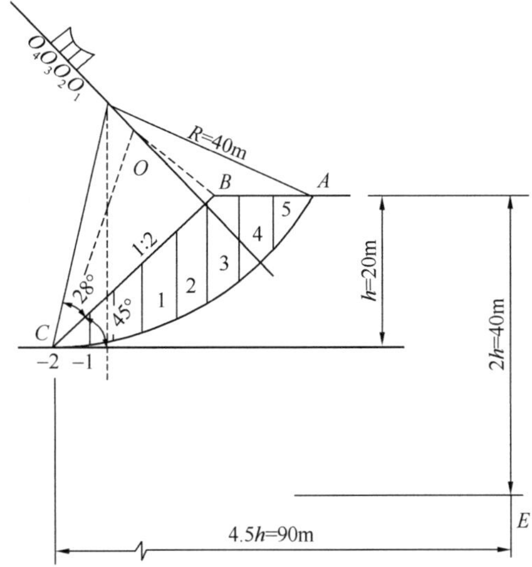

Figure 10. 8 Stability analysis of a slope by Swedish slice method

The calculation of Swedish slice method (Center number: O_1, H: 40m, Soil slice width: 8m)

Table 10. 2

Slice	h_i (m)	$\sin\theta_i$	$\cos\theta_i$	$h_i\sin\theta_i$	$h_i\cos\theta_i$
−2	3. 3	−0. 383	0. 924	−1. 26	3. 05
−1	9. 5	−0. 200	0. 980	−1. 90	9. 31
0	14. 6	0. 000	1. 000	0. 00	14. 60
1	17. 5	0. 200	0. 980	3. 50	17. 15
2	19. 0	0. 400	0. 916	7. 60	17. 40
3	17. 9	0. 600	0. 800	10. 20	13. 60
4	9. 0	0. 800	0. 600	7. 20	5. 40
sum				25. 34	80. 51

(4) Measure the center angle θ of the sliding arc as 98° and calculate the length of the sliding arc

（4）测量滑动弧的圆心角 θ 为 98°，计算滑动弧的长度。

$$\hat{L} = \frac{\pi}{180} \times \theta \times R = \frac{\pi}{180} \times 98 \times 40 = 68.4\text{m}$$

(5) Calculate the safety factor according to equation (10. 16) .

（5）根据式（10. 16）计算安全系数。

$$K=\frac{\gamma b\tan\varphi\sum h_i\cos\theta_i+c\hat{L}}{\gamma b\sum h_i\sin\theta_i}=\frac{10\times 68.4+18\times 8\times 0.364\times 80.51}{18\times 8\times 25.34}=1.34$$

(6) Reselect the center of the sliding arc O_2, O_3... on the EO extension line and repeat the above calculation to obtain the minimum safety factor, which is the stability safety factor of the soil slope.

（6）重新选择 EO 延长线上滑动弧 O_2、O_3……的圆心，重复上述计算，得到最小安全系数，即土坡稳定安全系数。

10.4.3 Bishop Slice Method

10.4.3 毕肖普条分法

In 1955, Bishop further developed the slice method. Assuming that the anti-sliding safety factor on the sliding surface at the bottom of each soil slice is the same, which is equal to the average safety factor of the entire sliding surface, calculate the unit length of the soil slope according to the plane problem, as shown in Figure 10.9. Assuming the possible sliding surface is a circular arc AC with a center of O and a radius of R. Divide the sliding soil ABC into several soil slices, and take any one of them (the i-th slice) to analyze its stress situation. The forces acting on the soil slice are: ① The self-weight of the soil slice is equal to its own weight, where b_i and h_i are the width and average height of the soil slice, respectively; ② The anti-shear force T_{fi}, effective normal reaction force N'_i, and pore water pressure $u_i l_i$ acting on the bottom surface of the soil slice, where l_i is the pore water pressure and sliding arc length at the midpoint of the bottom surface of the soil slice, respectively; ③ The normal forces E_i and E_{i+1} and tangential forces X_i and X_{i+1} act on both sides of the soil slice, $\Delta X_i=X_{i+1}-X_i$. And the points of action of G_i, T_{fi}, N'_i and $u_i l_i$ are all located at the midpoint of the bottom surface of the soil slice.

1955 年，毕肖普进一步发展了条分法，假设每个土条底部滑动面上的抗滑安全系数相同，等于整个滑动面的平均安全系数，根据平面问题计算土坡单位长度。如图 10.9 所示，假设可能的滑动面为圆弧 AC，圆心为 O，半径为 R。将滑动土 ABC 分成若干土条，取其中任意一条（第 i 条）分析其受力情况。作用在土条上的力为：①土条自重即土条自身的重量，其中 b_i 和 h_i 分别为土条的宽度和平均高度；②作用于土条底面的抗剪力 T_{fi}、有效法向反力 N'_i 和孔隙水压力 $u_i l_i$，其中 l_i 分别为土条底面中点处的孔隙水压力和滑动弧长；③法向力 E_i、E_{i+1} 和切向力 X_i、X_{i+1} 作用于土条两侧，$\Delta X_i=X_{i+1}-X_i$。G_i、T_{fi}、N'_i 和 $u_i l_i$ 的作用点均位于土条底面的中点。

According to the balance of the vertical force of the i-th soil slice, it can be obtained that

根据第 i 土条竖向力的平衡，可以得到式（10.19），

$$G_i+\Delta X_i-T_{fi}\sin\theta_i-N'_i\cos\theta_i-u_i l_i\cos\theta_i=0 \tag{10.19}$$

Further simplification can be written as

进一步简化为式（10.20）。

$$N'_i\cos\theta_i=G_i+\Delta X_i-T_{fi}\sin\theta_i-u_i b_i \tag{10.20}$$

When the soil slope is not yet damaged, the shear strength on the sliding surface of the soil slice is only

当土坡尚未破坏时，土条滑动面上的抗剪强度仅部分发挥作用。

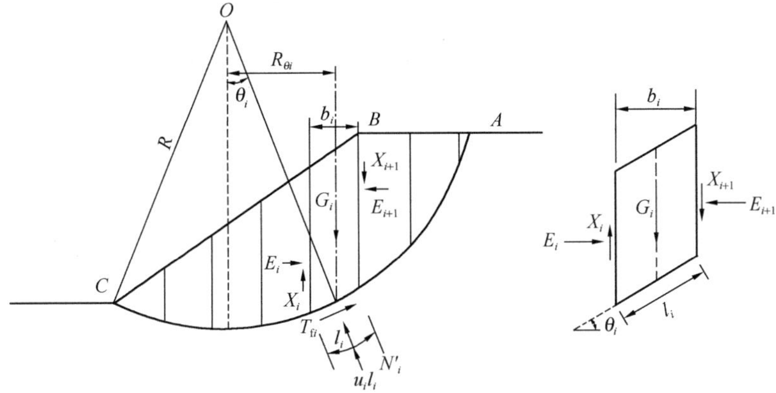

Figure 10.9 Stability analysis of a slope by Bishop slice method

partially exerted. If expressed as effective stress, the shear strength on the sliding surface of the soil slice is

如果用有效应力表示，则土条滑动面的抗剪强度为式（10.21），

$$T_{fi} = \frac{1}{K}(c'l_i + N'_i \tan\varphi') \tag{10.21}$$

where c', φ' are the effective stress intensity index of soil; K is the safety factor.

Substituting Eq. (10.21) into Eq. (10.20) yields

式中，c'，φ'为土的有效应力强度指标；K 是安全系数。将式（10.21）代入式（10.20）可得式（10.22），

$$N'_i = \frac{1}{m_{\theta i}}(G_i + \Delta X_i - u_i b_i - \frac{c'l_i}{K}\sin\theta_i) \tag{10.22}$$

in which

其中 $m_{\theta i}$可用式（10.23）表示。

$$m_{\theta i} = \cos\theta_i + \frac{1}{K}\sin\theta_i \tan\varphi' \tag{10.23}$$

Then, based on the moment balance of the entire sliding soil on the center O, the moments of the side wall forces between adjacent soil slices will cancel each other out, and the action lines of N'_i and $u_i l_i$ of each soil slice will pass through the center. Therefore, there is

基于整个滑动土在中心 O 上的力矩平衡，相邻土条之间的侧壁力矩将相互抵消，每个土条的 N'_i和 $u_i l_i$作用线将穿过中心，则有式（10.24），

$$\sum_{i=1}^{n} G_i R \sin\theta_i - \sum_{i=1}^{n} T_{fi} R = 0 \tag{10.24}$$

where which n is the number of slices, the other symbols are the same as before.

式中，n 是土条的数量，其他符号同前文。

Substituting Eqs. (10.22) and (10.24) into Eq. (10.21), and $b=b_i=l_i\cos\theta_i$, we can derive the expression of the safety factor as

将式（10.22）和式（10.24）代入式（10.21），且 $b=b_i=l_i\cos\theta_i$，可得安全系数的表达式为式（10.25）。

$$K=\frac{\sum \frac{1}{m_{\theta i}}[c'b+(G_i-u_ib+\Delta X_i)\tan\varphi']}{\sum G_i\sin\theta_i} \tag{10.25}$$

This is a general equation for calculating the safety factor of soil slopes using the Bishop's method, but ΔX_i is still unknown. In order to calculate K, it is necessary to estimate the value of ΔX_i, which can be solved by the successive approximation method. The trial values of X_i and E_i should satisfy the equilibrium conditions of each soil slice, and the $\sum \Delta X_i$ and $\sum \Delta E_i$ of the entire sliding soil are equal to zero. Bishop proved that if the ΔX_i of each soil slice is set to 0, the error generated is only 1%, which leads to the widely used Bishop simplified equation as

这是用毕肖普法计算土质边坡安全系数的一般公式，但 ΔX_i 仍然未知。为了计算 K，需要估计 ΔX_i的值，可以用逐次逼近法求解。X_i 和 E_i 的试验值应满足各土条的平衡条件，且整个滑动土的 $\sum \Delta X_i$ 和 $\sum \Delta E_i$ 均为零。毕肖普证明，如果将每个土条的 ΔX_i设为 0，则产生的误差仅为 1%，从而得到广泛使用的毕肖普简化方程为式（10.26）。

$$K=\frac{\sum \frac{1}{m_{\theta i}}[c'b+(G_i-u_ib)\tan\varphi']}{\sum G_i\sin\theta_i} \tag{10.26}$$

Due to the presence of a safety factor K in the calculation formula for $m_{\theta i}$ in Eq. (10.20), the above safety factor K still needs to be calculated through trial. Usually, during the trial calculation, $K=1$ can be assumed first, and $m_{\theta i}$ can be calculated. Then, K can be calculated according to Eq. (10.26). If the calculated K value is not the same as the assumed K value, the new $m_{\theta i}$ and K can be calculated by substituting the calculated K value. This process is repeated until the required accuracy is met for both previous and subsequent K values. Usually, 3~4 iterations are sufficient to meet the engineering accuracy requirements, and the iterations always converge.

由于式（10.20）中 $m_{\theta i}$ 的计算公式中存在一个安全系数 K，因此上述安全系数 K 仍需通过试验计算，在试验计算时，可先假定 $K=1$，然后计算出 $m_{\theta i}$，则可根据式（10.26）计算 K。如果计算出的 K 值与假设的 K 值不相同，则可以通过代入计算出的 K 值来计算新的 $m_{\theta i}$ 和 K。重复这个过程，直到前面和后面的 K 值都满足所需的精度，通常 3~4 次迭代就足以满足工程精度要求，并且迭代总是收敛的。

It should be noted that when θ_i is negative, $m_{\theta i}$ may approach zero, and N'_i will approach infinity, which is obviously unreasonable. Therefore, the simplified Bishop's slice method cannot be applied at this time.

需要注意的是，当 θ_i 为负时，$m_{\theta i}$可能会接近零，而 N'_i将接近无穷大，这显然是不合理的。因此，目前无法应用简化的毕肖普条分法。

To obtain the minimum safety factor K, it is necessary to assume several sliding surfaces and use the above method to calculate the safety factor K corresponding to each sliding surface. The minimum safety factor can be found as the safety factor of the slope,

为了获得最小安全系数 K，有必要假设几个滑动面，并使用上述方法计算每个滑动面对应的

and the corresponding sliding surface is the most dangerous sliding surface.

安全系数 K。最小安全系数可以作为边坡的安全系数，相应的滑动面是最危险的滑动面。

Problems

10.1 For the infinite slope shown in Figure 10.10 (consider that there is no seepage through the soil), determine:

a. The factor of safety against sliding along the soil-rock interface.

b. The height, H, that will give a factor of safety (F_s) of 2 against sliding along the soil-rock interface.

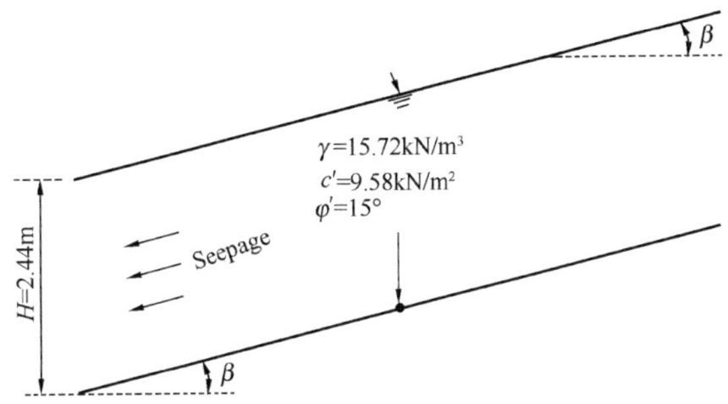

Figure 10.10 Figure of problem 10.1

10.2 Refer to Figure 10.10. If there is seepage through the soil as shown and the groundwater table coincides with the ground surface, what is the factor of safety, F_s, given H=1.16 m and γ_{sat}= 18.55kN/m³?

10.3 A cut slope is to be made in a soft saturated clay with its sides rising at an angle of 60° to the horizontal (Figure 10.11).

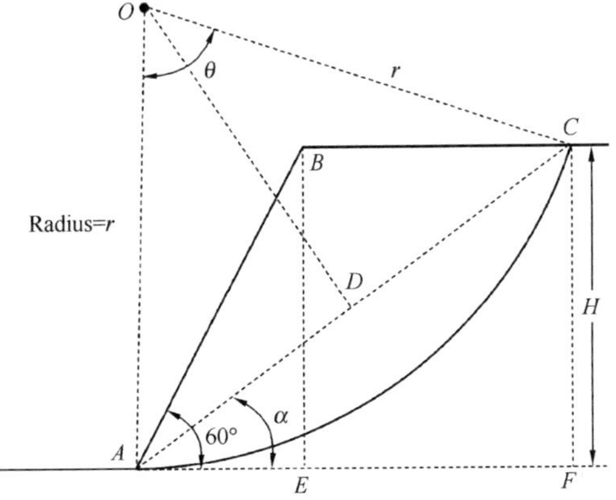

Figure 10.11 Figure of problem 10.3

Given: c_u= 40 kN/m² and γ=17.5kN/m³.

a. Determine the maximum depth up to which the excavation can be carried out.

b. Find the radius, r, of the critical circle when the factor of safely is equal to 1 (Part a).

c. Find the distance $\overline{BC}$.

参　考　文　献

［1］ 刘松玉．土力学[M]. 5 版．北京：中国建筑工业出版社，2020.

［2］ 李广信．高等土力学[M]. 2 版．北京：清华大学出版社，2016.

［3］ 中华人民共和国住房和城乡建设部. 土工试验方法标准：GB/T 50123—2019[S]. 北京：中国计划出版社，2019.

［4］ 中华人民共和国住房和城乡建设部. 岩土工程基本术语标准：GB/T 50279—2014[S]. 北京：中国计划出版社，2015.

［5］ Knappett J A，Craig R F. Craig's Soil Mechanics[M]. London：Spon Press，2012.

［6］ 中华人民共和国住房和城乡建设部. 建筑地基基础设计规范：GB 50007—2011[S]. 北京：中国计划出版社，2012.

［7］ 中华人民共和国建设部. 岩土工程勘察规范(2009 年版)：GB 50021—2001[S]. 北京：中国建筑工业出版社，2009.

［8］ 中华人民共和国建设部. 土的工程分类标准：GB/T 50145—2007[S]. 北京：中国计划出版社，2008.

［9］ 高大钊，袁聚云. 土质学与土力学[M]. 北京：人民交通出版社，2006.

［10］ 顾晓鲁，钱鸿缙，刘惠珊，等．地基与基础[M]. 3 版. 北京：中国建筑工业出版社，2003.

［11］ Das B M. Principles of Geotechnical Engineering[M]. Singapore：Cengage Learning，2002.

［12］ 龚晓南. 土力学[M]. 北京：中国建筑工业出版社，2002.

［13］ 赵明华. 土力学与基础工程[M]. 武汉：武汉工业大学出版社，2000.

［14］ 陈希哲. 土力学地基基础[M]. 3 版. 北京：清华大学出版社，1998.

［15］ 王滢，刘嘉怡，高盟，等. 地震作用下含气能源土动力特性试验研究[J]. 岩土力学，2025，46(2)：457-466.

［16］ 孔凡玲，王滢，孔祥霄，等. 深海能源土含气储层力学特性三轴试验研究[J]. 海洋工程，2023，41(6)：148-157.

［17］ 王滢，李朋飞，孔祥霄. 循环荷载下深海能源含气土动力特性试验研究[J]. 山东科技大学学报，2024，43(2)：30-39.